# Human Factors in Consumer Products

EDITED BY

## NEVILLE STANTON

UNIVERSITY OF SOUTHAMPTON

D1152946

Taylor & Francis
*Publishers since 1798*

UK    Taylor & Francis Ltd, 1 Gunpowder Square, London EC4A 3DE
USA  Taylor & Francis Inc., 1900 Frost Road, Suite 101, Bristol, PA 19007

Copyright © Taylor & Francis Ltd 1998

**British Library Cataloguing-in-Publication Data**

A catalogue record for this book is available from the British Library.
ISBN 0 7484 0603 4 (PB)
ISBN 0 7484 0602 6 (HB)

**Library of Congress Cataloguing-in-Publication Data are available**

Cover picture by Bryan Newman
Cover design by Amanda Barragry
Typeset in Times 10/12pt by MCS Ltd, Salisbury, UK
Printed by T.J. International Ltd, Padstow, UK

# Contents

    *Rachel Benedyk and Sarah Minister*

    3.1   The legal requirement for evaluation of product safety     55
    3.2   Formation of accidents                                     56
    3.3   The BeSafe method for safety evaluation                    58
    3.4   Applying BeSafe to product safety evaluation               60
    3.5   Limitations of BeSafe                                      63
    3.6   Case study: playground slides                              66
          Acknowledgement                                            72
          References                                                 73

4   **A systems analysis of consumer products**                     75
    *Neville Stanton and Christopher Baber*

    4.1   Predicting human error                                     75
    4.2   A systems approach                                         76
    4.3   Rewritable routines                                        79
    4.4   Task analysis for error identification                     83
    4.5   Conclusions                                                88
          References                                                 89

5   **Ergonomics and the evaluation of consumer products:**         91
    **surveys of evaluation practices**
    *Christopher Baber and Muna G. Mirza*

    5.1   Introduction                                               91
    5.2   Survey One: technical testing and evaluation of white goods  94
    5.3   Survey Two: ergonomics evaluation practice                 98
    5.4   Discussion                                                 102
          References                                                 103

PART TWO   **Institutions involved in design and evaluation of**    105
           **consumer products**

6   **Application of ergonomics and consumer feedback to**          107
    **product design at Whirlpool**
    *Adrian Martel*

    6.1   Introduction                                               107
    6.2   High-profile ergonomic development                        107
    6.3   Interaction design                                        114
    6.4   Low-profile ergonomic development                         122
    6.5   Conclusions                                               125
          References                                                 126

# Contributors

Christopher Baber
Industrial and Ergonomics Group
School of Manufacturing and Mechanical Engineering
University of Birmingham
Birmingham
B15 2TT
UK

Rachel Benedyk
Ergonomics and HCI Unit
University College London
26 Bedford Way
London
WC1H 0AP
UK

John V.H. Bonner
Institute of Design
Teesside University
Middlesborough
TS1 3BA
UK

Lindsey M. Butters
Consumers' Association Research and Testing Centre
Davy Avenue
Knowlhill
Milton Keynes
MK5 8NL
UK

Mark Evans
Department of Design and Technology
Loughborough University
Loughborough
Leicestershire
LE11 3TU
UK

Alan Hedge
Department of Design and Environmental Analysis
Cornell University
MVR Hall
Ithaca
NY 14853
USA

Patrick W. Jordan
Philips Corporate Design
Building W
Damsterdiep 267
PO Box 225
9700 AE Groningen
THE NETHERLANDS

Alastair S. Macdonald
Product Design Engineering
Glasgow School of Art
167 Renfrew Street
Glasgow
G3 6RQ
UK

Adrian Martel
Manager, Centre for Applied Product Ergonomics
Central Industrial Design
Whirlpool Europe s.r.l.
Biandronno (VA)
ITALY

Sarah Minister
Ergonomics and HCI Unit
University College London
26 Bedford Way
London
WC1H 0AP
UK

Muna G. Mirza
Industrial and Ergonomics Group
School of Manufacturing and Mechanical Engineering
University of Birmingham
Birmingham
B15 2TT
UK

Magdalen Page
ICE Ergonomics
Holywell Building
Holywell Way
Loughborough
Leicestershire
LE11 3UZ
UK

Neville Stanton
Department of Psychology
University of Southampton
Highfield
Southampton
SO17 1BJ
UK

Bronwen Taylor
Philips Corporate Design
Philips International BV
Building OAN 3
PO Box 218
5600 MD Eindhoven
THE NETHERLANDS

Bruce Thomas
Philips Corporate Design
Philips International BV
Building OAN 3
PO Box 218
5600 MD Eindhoven
THE NETHERLANDS

Monica Trommelen
Centre for Safety Research
Faculty of Social Sciences
Leiden University
PO Box 9555
2300 RB Leiden
THE NETHERLANDS

Mark Young
Department of Psychology
University of Southampton
Highfield
Southampton
SO17 1BJ
UK

Harm J. Zwaga
Department of Psychonomics
Faculty of Social Sciences
Utrecht University
THE NETHERLANDS

# Foreword

There is a saying in the design profession – 'If you're not part of the solution, you're part of the problem'. This is never more true than when applied to the design of products. Designers have the potential to be part of the solution. Too often in the past, designers have simply exacerbated the problem – designing furniture, products and space-planning schemes that make matters worse.

We all know the role that good ergonomics can play in creating a better, healthier living and working space. I would say that good design and good ergonomics go hand in hand. Ergonomics is an essential factor in making the broader contribution of design solutions in our lives work more effectively.

But then perhaps you would expect me to say this – as Design Director at the Design Council, an organisation dedicated to improving the prosperity and well-being of the UK through inspiring the best use of design.

I welcome this book and hope it provides valuable insights into how designers, engineers and ergonomists can work together in designing a better world to live and work in. Only by adopting this team approach, and with access to the latest knowledge can we expect businesses to embrace best practice and compete successfully in world markets.

As living and working patterns change, through the enabling potential of technology, we need, more than ever, well designed products that improve the quality of life as well as the prosperity of those manufacturing companies. Best practice in product design is all too rare. Whilst notable exceptions amongst companies design delightfully refined, well conceived products that do actually improve the way we live and work, there are still many infuriating products that fill our lives with frustration.

It is a credit to Neville Stanton and the contributors that they have undertaken to write this book. I wish it well.

Sean Blair
Design Director
Design Council

# Preface

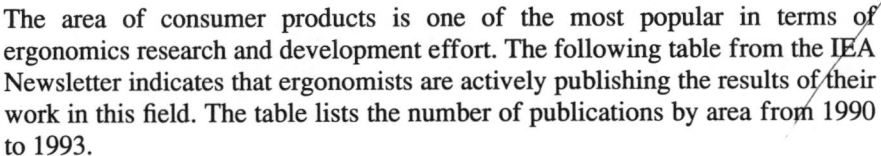

The area of consumer products is one of the most popular in terms of ergonomics research and development effort. The following table from the IEA Newsletter indicates that ergonomists are actively publishing the results of their work in this field. The table lists the number of publications by area from 1990 to 1993.

| | |
|---|---|
| 4500 | Consumer products |
| 4200 | Safety and health |
| 2700 | Human/computer interaction |
| 2300 | Strain and psychophysiology |
| 1400 | Control rooms and human reliability |
| 1300 | Organisational design and management |
| 1200 | Musculoskeletal disorders |
| 1000 | Human aspects of advanced manufacturing |
| 900 | Aging |
| 900 | Cost-effective ergonomics |
| 900 | Standards |
| 600 | Rehabilitation |
| 500 | Agriculture |
| 400 | Building and architecture |
| 200 | Work environment design |
| 50 | Industrial design |
| 50 | Quality |

Every consumer product is designed. The extent to which products are designed for use by people is the main question addressed by this book. Human factors (cf. ergonomics) offer a unique perspective on design, as it is concerned with the interaction between the consumer and the product with a view to optimising performance and pleasure. Consumer products are so much part of our everyday life that we probably give little thought to their design, except when we experience difficulties. The chapters within this book provide examples of the design process, methods for design, design case studies and guidelines. The idea for this book came to me after I had organised a one-day symposium on 'human factors in consumer product design and evaluation' on behalf of the Ergonomics Society of the UK, with co-sponsorship from the Consumers' Association, held at the University of Southampton in November 1995. The event was well attended and there was a large number of requests for copies of the proceedings after the event. This book has benefited from many additional contributions and substantial revision of the original conference papers.

I would like to thank everyone who has made this book possible, including all of the contributors, the staff at Taylor & Francis (particularly Richard Steele, Andrew Carrick and Mariangela Palazzi-Williams), Steve Konz for supplying the list of Ergonomics contacts, Sean Blair (Design Council) for writing the Foreword and Dr Deborah Brown (Consumers' Association) for her help in running the conference. Special thanks go to Maggie Stanton, Joshua and Jemima, for enduring my absence during this and other projects which have kept me away from my role as husband and father.

Neville Stanton
*University of Southampton*

# CHAPTER ONE

# Product design with people in mind

NEVILLE STANTON

*Department of Psychology, University of Southampton*

## 1.1 What are Human Factors and Ergonomics?

Engineering is concerned with improving products from the point of view of mechanical and electrical design, and psychology is concerned with the study of the mind and behaviour. Human factors and ergonomics are concerned with adapting products to people, based upon their physiological and psychological capacities and limitations (Blum, 1952), the objective being to improve overall system performance (involving human and product elements). As Sanders and McCormick (1987) put it, 'it is easier to bend metal than twist arms', by which they mean that the design of the device to prevent errors is likely to be more successful than telling people not to make errors when using a device. The overall objectives of ergonomics and human factors are to optimise the effectiveness and efficiency with which human activities are conducted as well as to improve the general quality of life through 'increased safety, reduced fatigue and stress, increased comfort [and] ... satisfaction.' (Sanders and McCormick, 1992, p. 4). It is difficult to delineate the genesis of human factors and ergonomics, but both can be traced back to a general interest in problems at munitions factories during the First World War (Oborne, 1982). Machines that were designed to be operated by men seemed to have production-related problems when operated by women. These difficulties were resolved when it was realised that the problems were related to equipment design rather than the people operating them, i.e. they were designed to be operated by men and not women. The misreading of altimeters by pilots in the Second World War stimulated further interest in human factors and ergonomics. A study by Grether (1949) illustrated that not only did pilots take over 7 seconds to interpret the traditional three-needle altimeter (where the three pointers read

1

tens of thousands, thousands and hundreds of feet respectively) but also that nearly 12 per cent of the readings contained errors of 1000 feet or more. Grether showed conclusively that superior designs could dramatically reduce both reading time and error rate. This study, perhaps more than any other, indicates the importance of psychology in the design of devices. Despite this evidence, it is sometimes difficult to gain acceptance from the engineering community and to change design, as the following quote from an accident report in 1958 (some nine years after Grether's original study) shows:

> The subsequent investigation ... showed that the captain had misread his altitude by 10,000 feet and had perpetuated his misreading error until the aircraft struck the ground and crashed. (Rolfe, 1969, p.16)

The terms 'human factors' and 'ergonomics' will be used interchangeably throughout this book. This is not meant to confuse the reader, rather it reflects that they are synonymous. The Human Factors Society in the USA has recently changed its name to The Human Factors and Ergonomics Society in recognition of this. Although the origin of the discipline is in workplace domains, the subject has expanded to cover all manner of human interaction with technological artefacts, including consumer products.

## 1.2  The Need for Human Factors

We are all familiar with the frustrations that accompany the use of technology in the home and at work. Norman (1988) provides an abundance of examples on this subject. His book begins with an anecdote about Kenneth Olsen (the engineer who founded and still runs Digital Equipment Corporation) who cannot heat a cup of coffee in his company's microwave oven. I have a similar story. I was recently invited to a symposium run by a major consumer products manufacturer, and to everyone's amusement the combined effort of the human factors group could not get its own video cassette recorder to play a video tape. Norman suggests that people have difficulties in operating all manner of consumer products: washing machines, dryers, telephones, televisions, stereos, VCRs, refrigerators and so on. Why do these devices, which are supposed to make our life easier, seem to thwart our best intentions? One reason is that users of these devices perceive the problem to be with themselves rather than with the technology. People often blame themselves when failing to comprehend the manufacturer's instructions or when errors occur (Reason, 1990). Also, the problems are usually of a fairly trivial and individual nature, and do not affect other people. These problems are often only minor hassles compared with major events, such as incidents in the aviation and nuclear industries. On the face of it there is little comparison between errors with VCRs and errors on the flightdeck of an aircraft. However, Reason (1990) argues that, at the basic level of interfacing human thought processes with technology, there are many similarities. Despite the obvious differences in training, level of skill and knowledge in operating VCRs and aircraft, basic error types such as mode error (i.e. errors that occur when devices have different modes of operation and the action appropriate for one mode has different consequences in other modes) have been found to occur in both environments (Norman, 1988).

## 1.3  Demand/Resource Theory in Human Factors

Solutions to the problems raised in people interacting with technology come in two main forms: either to reduce demand or to increase resources in situations of work overload (vice versa in situations of work underload). The dual concepts of demands and resources are prevalent in engineering psychology and particularly pertinent when considering the capacities and limitations of people in technological environments. Wickens (1992) proposes a theory of multiple pools of attentional resources in relation to different information processing demands – speech and text utilise a verbal information processing code and draw upon a different pool of attentional resources from tones and pictures which utilise a spatial processing code. Wickens argues that when the attentional resources assigned to the verbal processing code are exhausted, workload demands may be increased further by using the alternative spatial information processing code through the presentation of tones or pictures (although these pools are not mutually exclusive).

The concepts of *demands* and *resources* (see Figure 1.1) provide a conceptual framework for human factors. Demands and resources could come

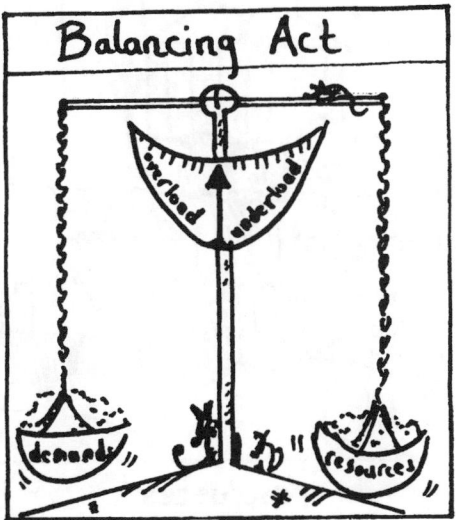

**Figure 1.1**  Demands and resources in human factors.

from the task, the device, the user and the environment. For example, user resources (e.g. knowledge, experience and expertise) and demands (e.g. user goals and standards) interact with task demands (e.g. task goals and standards) and task resources (e.g. instruction manuals and training). This interaction is mediated by demands (e.g. device complexity) and resources (e.g. clarity of the user interface, which could reduce demands) of the device being operated. This is a familiar concept in discussions of task workload, and it is implied that demand/resource imbalance can occur as either task underload or task overload, both of which are detrimental to performance. An illustration of the relationship between demands and resources is provided by the tale of Procrustes (Oborne, 1982). In Greek mythology, Procrustes was an ingenious robber who persuaded travellers to part with their gold. His trick was very simple. He offered weary travellers all the food and wine they wanted and they could either pay for what they had consumed (at a price of Procrustes' choice) or accept his hospitality without payment and take a bed for the night. Most travellers opted for the latter, at which point Procrustes added one more clause: that the traveller had to fit one of his two spare beds exactly. Most accepted without question and ate and drank their fill. When it came time for them to bed down for the night, Procrustes showed them the two beds, one was very long and the other very short. At this point Procrustes threatened to make them fit

**Figure 1.2**  The Procrustean approach.

the bed by either cutting off their legs to fit the short bed or stretching them to fit the long bed (Figure 1.2). Most travellers ended up paying the exorbitant fees for what they had consumed.

Oborne (1982) suggests that the Procrustean approach often appears to be (inadvertently) taken by designers, who design tasks that either stretch people beyond their physical and/or mental capacities or restrict them physically and/ or mentally. Both ends of the spectrum result in a dissatisfactory outcome for the user, who end up paying for poor design in terms of errors and dissatisfaction with the product.

## 1.4 Consumer Product Design

There is nothing new about the idea of designing products. People have been designing products for centuries. The Abbevillian hand axes and stone spear points crafted over 10,000 years ago must have been formed as a result of some image of the finished product in mind. Ultimately, a consensus was probably reached concerning an ideal form for each product for each task. This probably only occurred after much trial and error and successive refinement over many years. Human factors has become concerned with understanding the design process for two main reasons. First, there is a concern for optimising the design

**Christian Dell (1893-1974) German**

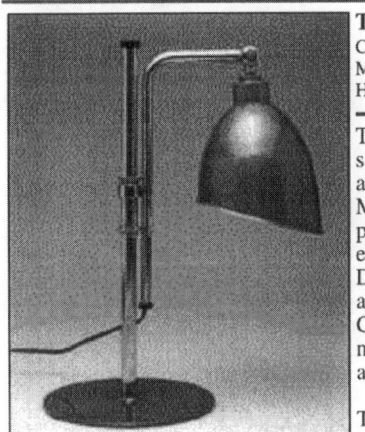

**Table Lamp**, 1928.
Copper shade, chromium-plated metal. Adjustable shade.
Manufacturer: Betmag, Zurich.
Height: 16 1/2" x Width: 9 1/2" // 92-12

Today, German Bauhaus design is best known for its simplicity of form, clean lines, inexpensive materials and, above all, for its functionalism. This was not always so. Many early works, especially those created in the Weimar phase (1919-23), were highly individualistic, certainly more expressionist in form and spirit. When the school moved to Dessau in 1925 Lazlo Moholy-Nagy (1895-1946) was appointed Form Master of the new metal workshops and Christian Dell the Master Craftsman. The two jointly forged a new design vocabulary for which Bauhaus metalwork is admired today.

This rare table lamp (produced after Dell left the Dessau Bauhaus in 1925) is a superb example of pure Bauhaus design; it is the only know example in this country. The shape of the modern office table lamp was largely developed by him, and his designs of the '20s, such as this lamp, were still being produced some three decades later. The simplified geometric form--totally absent of ornamentation--and mechanized appearance have become the hallmarks of mass-produced industrial design to this day.

**Figure 1.3** A table lamp (source: Norwest Corporation (Minneapolis) collection).

process, to reduce the effects of chance and errors in design. Secondly, there is the concern to incorporate the requirements of the end-user as early as possible, when design is relatively fluid. It is argued that this process is product-independent.

In perusal of the World Wide Web, I became interested in the common goals of the human factors and design community when I found a short article on the Bauhaus school of art founded by Walter Gropius. The Bauhaus movement is famed for the design philosophy of *form following function*. This seems to share the goals of human factors, which mediates form by consideration of the characteristics of the user population and the task. Figures 1.3 and 1.4 illustrate two consumer products from the Bauhaus tradition (this is not intended to represent the design community in which there are many schools of thought, but rather to offer a simple illustration of one movement). It is important for the human factors community to understand the design community and vice versa.

Human factors and design are from very different traditions, however.

 Marguerite Friedlander-Wildenhain (1896-1985) German

**Teapot**, 1930. Hard-paste porcelain. Manufacturer: German State Porcelain Factory, Berlin. Height: 5" x Width 11" x Depth: 7" // 89-75

The Bauhaus was unceremoniously closed in 1933 but its legacy--a rational approach to design--remained alive and a few industrial prototypes were put into production. These included pottery for the State Porcelain Factory in Berlin. Although relatively insignificant in terms of generated income and the wider context of manufacturing output, they are lasting reminders that the modernist industrial aesthetic of the Bauhaus and other progressive design centers of the 1920s were not entirely obliterated by the destructive tide of the Third Reich.

The soft mint-green glaze and clean surface of this teapot result in a design both elegant and modern. The teapot's form and function are superbly resolved in three essential areas: the spout pours easily without dripping (note the detail at the lip); the lid is countersunk so it won't fall out as the pot is tilted; and the large handle facilitates the pouring of tea with relative ease. Simplicity of form and timeless design well-adapted to industrial manufacture, these are the hallmarks of one of the finest designs emanating from the Bauhaus.

**Figure 1.4** A teapot (source: Norwest Corporation (Minneapolis) collection).

Whilst design emanates from the arts and crafts disciplines, human factors comes from the Sciences. Perhaps this, more than any other reason, is the cause for communication difficulties between the two groups. The human factors approach is based upon understanding user activity with products. A simple schematic of product user is shown in Figure 1.5.

Figure 1.5 illustrates a cycle of user activity, beginning with the goals of the user and moving through to forming intentions, planning and executing actions. These actions will lead to some change in the product. This change has to be perceived and interpreted by the user, then evaluated against the overall goal. The comparison with the user goals will inform the user if the desired change in product status has been achieved. This is normally in relation to some specific task, such as cooking a meal, making a cup of tea, recording a TV programme, washing some clothes, etc. Errors in the human/product activity cycle (shown in Figure 1.5) can occur at three points. First the product user can be *mistaken* in his or her perception, interpretation and evaluation the current state of the product. Secondly, the user could *forget* to incorporate an action in the planned sequence. Finally, the user could execute an action where another action was intended. By understanding the underlying cause of these errors, we may begin

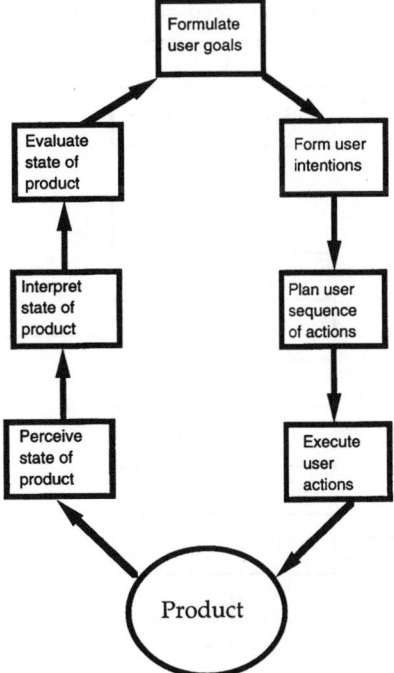

**Figure 1.5**  User activity with products (adapted from Norman, 1988).

to anticipate them and design products accordingly (Norman, 1988; Reason, 1990). Incorporation of the user requirements, user goals and user tasks into the design of a product will only be realised through a user-centred design process as illustrated in Figure 1.6.

Given that user activity is central to design, this information needs to be captured and incorporated into the design process. Design may start with a design brief specified by the client organisation. This will carry with it specific constraints (such as resource constraints in terms of time and money) as well as an outline functional specification (such as the range of functions that the product should perform). The user-centred approach would start the design activities with methods aimed at capturing user needs. These needs, together

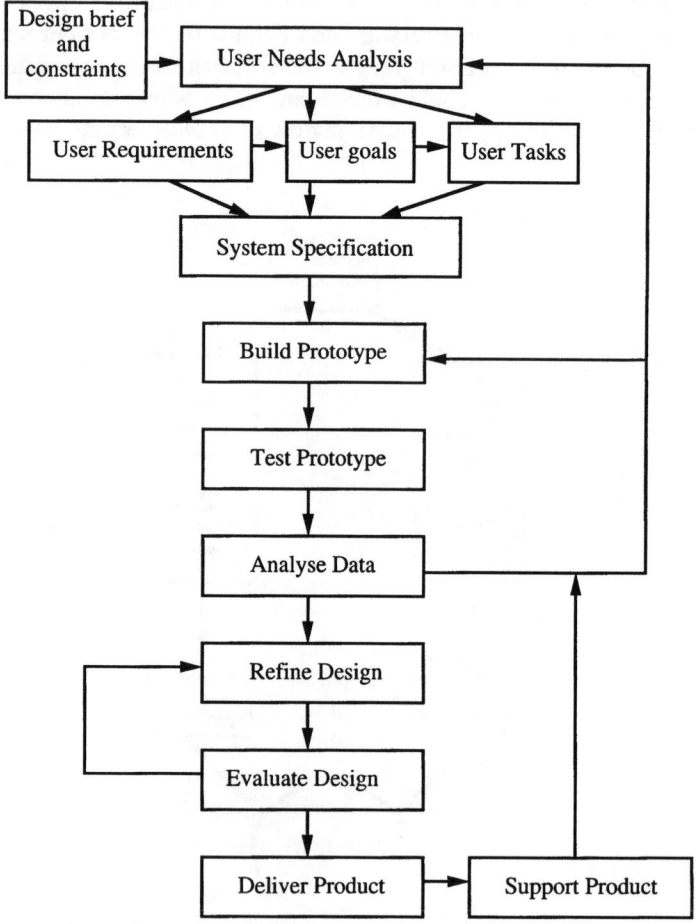

**Figure 1.6** A user-centred design process.

with the functional specification and the constraints will be formulated into the system specification. Additional constraints might include the cost and weight of materials and the size of components. From this a prototype may be built and tested. The resultant data may be used to inform the design team. This could lead to a complete overhaul of the system specification (reversing the design process back to a re-analysis of the user needs), or modification of the prototype, or minor refinements in the design. There may be many feedback loops around this process, or some stages may be omitted owing to resource constraints. Finally the product and associated materials (such as instructions, packaging and marketing) are delivered. Depending upon the life of the product, it may have to be supported by production of spares, maintenance contracts and helplines. Whilst this description represents a sanitised version of the design process, it does begin to touch on some of the complexity involved in design. This complexity is exacerbated when one appreciates the number of people who have a stake in product design, including the designers, ergonomists, accountants, management, technicians and engineers. If this group is not working harmoniously, agreement on the design solution is not likely to be forthcoming. At the early stages of design there may be as many as ten concepts for the product. This may lead to, say, three proposals which are narrowed down to one product. The process of freezing a range of concepts into a single solution makes it clear why human factors needs to get into the process as early as possible, as this is when most change can be made at least cost to the design process.

## 1.5 Usability in Product Design

While it is possible to indicate the necessity for usability in product development, as a concept it has proved remarkably resilient to definition. We all know what it is, but have difficulty reaching an agreed, coherent definition which will allow recommendations to be made concerning how best to make something more usable. Therefore the first, and perhaps most important, stumbling block in design is to agree upon what makes a product usable. If we cannot agree on what usability is, how can we hope to measure it? It is likely that different definitions of the concept will lead people to measure different aspects of product use. This suggests that a usability evaluation may not have a common standard between individuals. If usability is to be more than an ephemeral concept, we must agree on its constituent ingredients. Stanton and Baber (1992, 1996) draw upon a decade of work represented by Shackel (1981), Eason (1984) and Booth (1989) to suggest that the factors below serve to shape the concept of usability and define its scope:

- *Learnability* A system should allow users to reach acceptable performance levels within a specified time.

- *Effectiveness* Acceptable performance should be achieved by a defined proportion of the user population, over a specified range of tasks and in a specified range of environments.
- *Attitude* Acceptable performance should be achieved within acceptable human costs, in terms of fatigue, stress, frustration, discomfort and satisfaction.
- *Flexibility* The product should be able to deal with a range of tasks beyond those first specified.
- *The perceived usefulness or utility of the product* Eason (1984) has argued that the major indicator of usability is whether a product is used. As Booth (1989) points out, it may be possible to design a product which rates high on the LEAF (learnability, effectiveness, attitude, flexibility) precepts, but which is simply not used.
- *Task match* In addition to the LEAF precepts set out above, a usable product should exhibit an acceptable match between the functions provided by the system and the needs and requirements of the user.
- *Task characteristics* The frequency with which a task can be performed and the degree to which the ask can be modified, e.g. in terms of variability of information requirements.

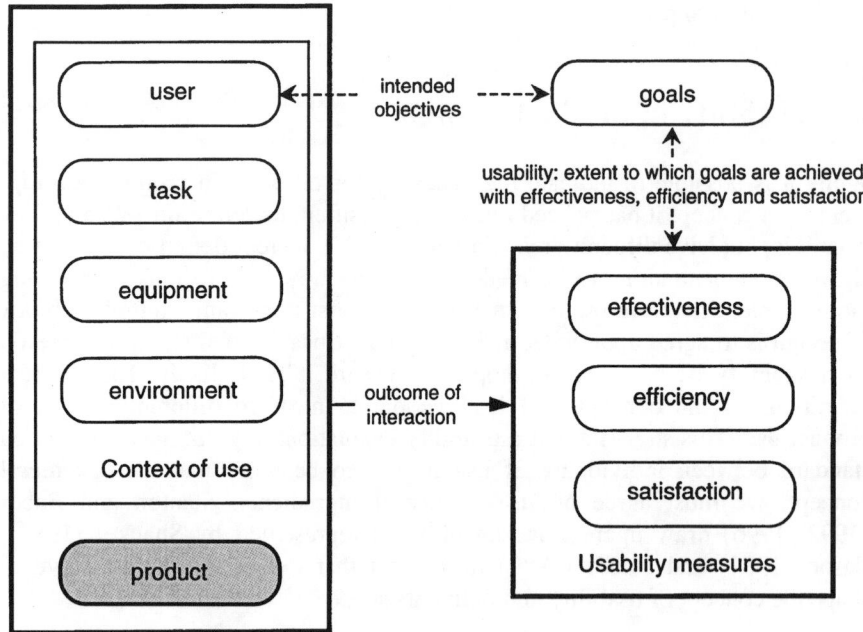

**Figure 1.7**  Usability framework from ISO 9241.

■ *User characteristics* Another section which should be included in a definition of usability concerns the knowledge, skills and motivation of the user population.

The ISO 9241 definition of usability is:

> the extent to which a product can be used by specified users to achieve specified goals with effectiveness, efficiency and satisfaction in a specified context of use.

This definition was intended for the design of software products but it is sufficiently generic to apply to any kind of product. The definition seems to incorporate many of the concepts identified by Stanton and Baber (1992, 1996). Within ISO 9241, a usability framework is proffered, as illustrated in Figure 1.7.

The measurement criteria are based upon Shackel's LEAF precepts, except that the flexibility criterion appears missing. Task match, task characteristics and user characteristics are implied. There is little evidence of the perceived usefulness or utility of the product present in the criteria. Despite any shortcomings, the framework is likely to serve the product designer well in embarking upon investigations into product usability. There is likely to be increased representation of human factors and ergonomics in consumer product standards as awareness in the discipline grows.

## 1.6 Standards for Consumer Products

Under the Sale of Goods Act 1979 (as amended by the Sale and Supply of Goods Act 1994 and the Sale of Goods (Amendment) Act 1994), consumer products in the UK should be 'fit for the purpose for which they were intended' and should 'correspond with the description of them'. This law aims to protect the consumer from poor and faulty goods. The interested reader is referred to the literature on mercantile law to pursue this topic, but it is just worth noting that there is legislation to protect the consumer. Unlike legislative regulations, standards (such as those produced by the British Standards Institution (BSI), European Committee for Standardisation (CEN) and the International Organisation for Standardisation (ISO)) tend to be voluntary. A standard is defined as:

> A document, established by consensus and approved by a recognised body that provides, for common and repeated use, rules, guidelines or characteristics for activities or their results, aimed at the achievement of the optimum degree of · order in a given context. (ISO/IEC Guide 2)

An illustration of the standardisation processes for BSI, CEN and ISO is set out in Figures 1.8, 1.9 and 1.10, respectively.

All countries have their own national standards body. The BSI is the independent body responsible for preparing British standards (tel.:

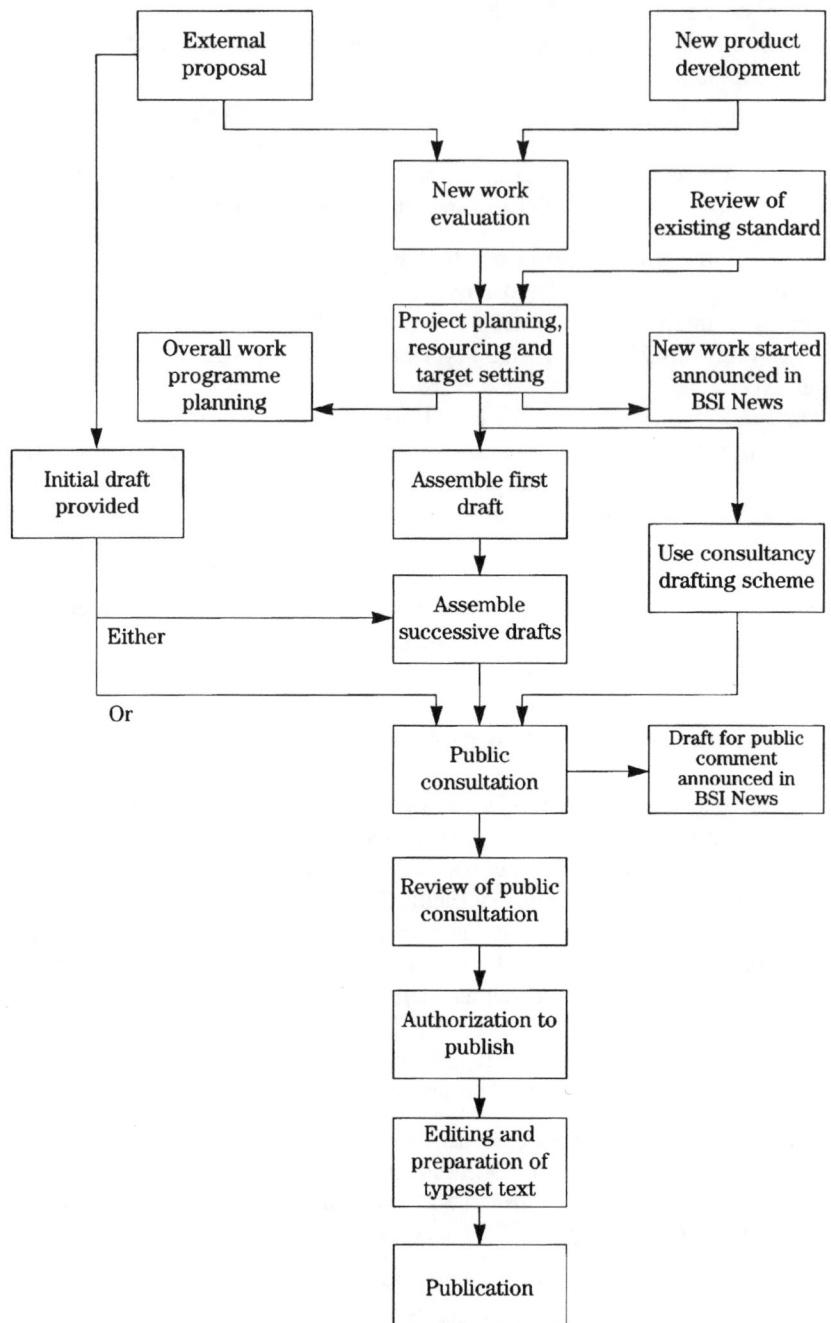

**Figure 1.8**  Procedure for the development of a BSI standard.

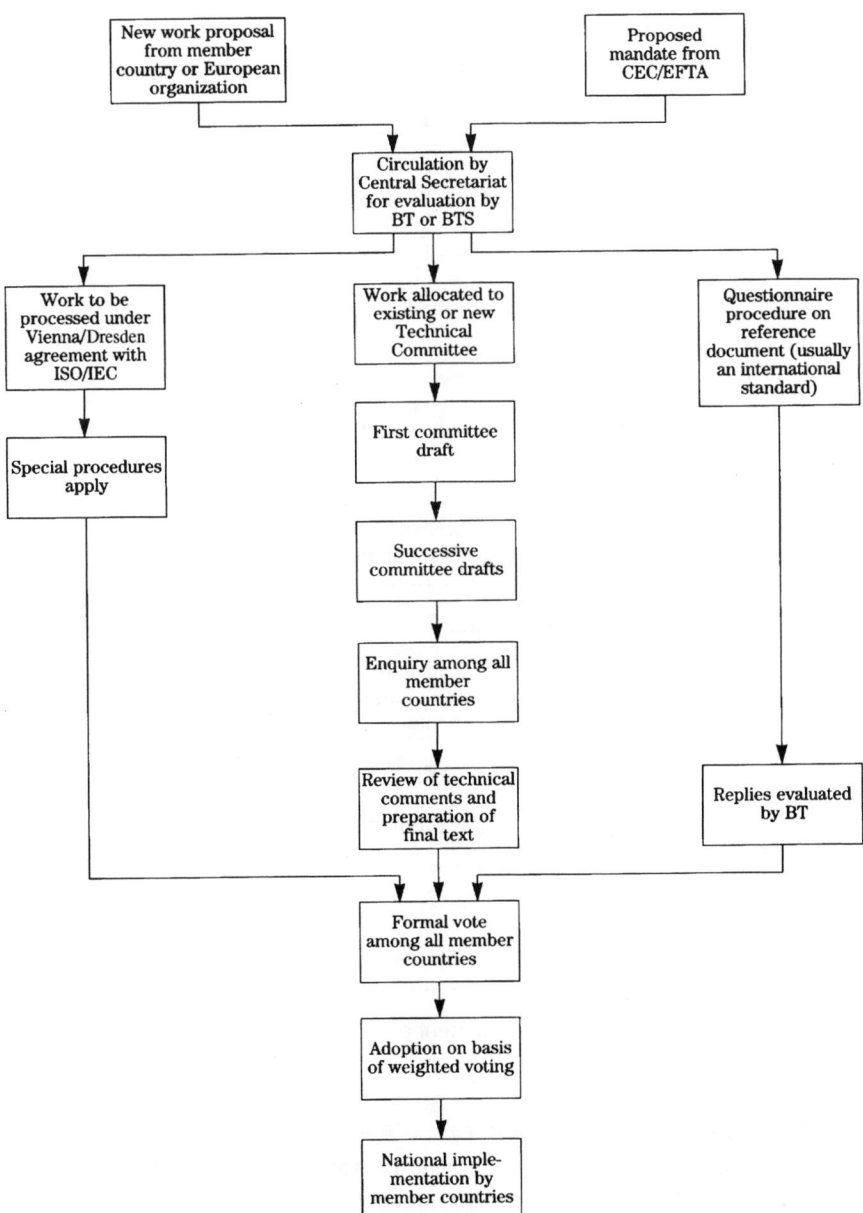

**Figure 1.9** Procedure for the development of a CEN standard.

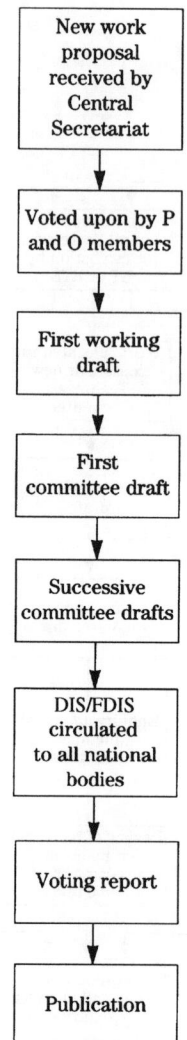

**Figure 1.10** Procedure for the development of an ISO standard.

0181–996 7000; fax: 0181–996 7001). British standards define safety and quality requirements for an enormous range of consumer products, including electrical and gas appliances, furniture, toys and leisure equipment. For example, BS 3456 (1987) is concerned with the safety of household and similar electrical appliances. Part of the standard is reproduced below to illustrate that it covers both abnormal and normal product use.

### BS 3456 (1997) Safety of Household and Similar Electrical Appliances

Part 101: General Requirements

19.1 Appliances shall be designed so that the risk of fire and mechanical damage impairing safety or protection against electrical shock as a result of **abnormal or careless use** is obviated as far as possible.

20.2 Moving parts of motor-operated appliances shall, as far as is compatible with the use and working of the appliance, be so arranged or enclosed as to provide, **in normal use**, adequate protection against personal injury.

The focus on use of the product would seem to indicate a clear role for the human factors professional in product design and development in order that the final version of the product entering the marketplace may be optimised to meet the requirements of the standard.

All standards are periodically reviewed, which may lead to calls for changes in existing standards. New standards may result from new product development or from a call for a standard by a body such as the Consumer Policy Committee (CPC) which is staffed by volunteer consumer representatives from organisations such as the Consumers' Association, the Department of Trade and Industry, the Institute of Consumer Affairs, the Officer of Fair Trading and the Royal Society for the Prevention of Accidents. The membership reflects an interest in protecting the consumers' interests and the aims of the CPC are to:

- improve consumer safety and health;
- enhance the performance of consumer products;
- improve information for consumers on products;
- facilitate consumer choice;
- contribute to environmental protection.

The consumer representation is to ensure that consumers' expectations of the safety and quality of consumer products are properly addressed when considering standards. The effectiveness of product information is increasingly being seen as an important role for standardisation (e.g. product labelling and instructions for product use) to ensure that any symbols or pictograms used are readily understood by the consumer. Unless made otherwise, standards are voluntary, and the onus is on the user to ensure that the standard is appropriate. A standard becomes binding for any consumer product if it has been made mandatory by legislation or a claim of compliance has been made. The CPC is represented on approximately 300 standards-making committees in the technical subject areas shown in Table 1.1. There are far too many standards associated with consumer products for them all to be mentioned, but the reader is referred to the BSI Standards Catalogue which is kept by most large libraries and is also available from the British Standards Institution.

**Table 1.1**   Range of technical areas covered by the Consumer Policy Committee

| Electrical, electronic and other fuel-using equipment (OC/11/9) | Household products, furniture and textiles (OC/11/10) | Mechanical engineering construction and health care (OC/11/11) | Personal protection systems and services (OC/11/12) |
| --- | --- | --- | --- |
| Household electrical appliances | Cooking and table utensils | Caravans | Car safety and security |
| Electrical installation | Furniture | Machinery safety and ergonomics | Alarm systems |
| Powered tools, DIY equipment and garden machinery | Child care articles and nursery goods | Tools | Protective clothing |
| | Paper and packaging | Sanitary appliances | Sports and leisure equipment |
| Lamps and lighting | Floor coverings | Stairs, doors and windows | Graphical symbols and signs |
| Gas appliances | Wall coverings | Fire safety | Fireworks, matches and lighters |
| Solid-fuel appliances and barbecues | Clothing and footwear | Contraceptives | Environmental issues |
| Solar heating | Bedding | Aids for disabled people | |
| Road transport informatics | Toys | Building products | Banking, smart cards and information technology |
| Metering | | Spectacles and contact lenses | Quality management |
| | | | Services |

## 1.7   Structure of the Book

The book is structured into three parts: methods in product design, institutions involved in design and evaluation of consumer products, and guidance and examples of product design. Part One contains chapters on comparing a range of methods (chapter 2), the BeSafe method (chapter 3), the TAFEI method (chapter 4) and a survey of current practice in industry (chapter 5). In Part Two there are chapters from Whirlpool (chapter 6), ICE (chapter 7), Philips (chapter 8) and the Consumers' Association (chapter 9). In the final part, Part Three, there are chapters on aesthetics (chapter 10), the development of a strimmer (chapter 11), hand-operated devices (chapter 12), the development of symbols for product warnings (chapter 13) and some general product development guidelines (chapter 14). The concluding chapter summarises the contributions and offers an analysis of the key human factors issues in consumer product design.

## References

BLUM, M. L. (1952) *Readings in Experimental Industrial Psychology*. New York: Prentice Hall.

BOOTH, P. (1989) *An Introduction to Human Computer Interaction*. London: LEA.

EASON, K. D. (1984) Towards the experimental study of usability. *Behaviour and Information Technology* **3**(2), 133–43.

GRETHER, W. F. (1949) Instrument reading. 1: The design of long-scale indicators for speed and accuracy of quantitative readings. *Journal of Applied Psychology*, **33**, 363–72.

NORMAN, D. A. (1988) *The Psychology of Everyday Things*. New York: Basic Books.

OBORNE, D. J. (1982) *Ergonomics at Work*. Chichester: Wiley.

REASON, J. (1990) *Human Error*. Cambridge: Cambridge University Press.

ROLFE, J. M. (1969) Human factors and the display of height information. *Applied Ergonomics*, **1**, 16–24.

SANDERS, M. S. and MCCORMICK, E. J. (1987) *Human Factors in Engineering and Design*. New York: McGraw-Hill.

SANDERS, M. S. and MCCORMICK, E. J. (1992) *Human Factors in Engineering and Design* (2nd edition). New York: McGraw-Hill.

SHACKEL, B. (1981) The concept of usability. *Proceedings of the IBM Software and Information Usability Symposium*. New York: IBM, pp. 1–30.

STANTON, N. A. and BABER, C. (1992) Usability: concept, practice and its relationship with EC Directive 90/270/EEC. *Displays*, **13**(3), 151–60.

STANTON, N. A. and BABER, C. (1996) Factors affecting the selection of methods and techniques prior to conducting a usability evaluation. In Jordan, P. W., Thomas, B., Weerdmeester, B. A. and McClelland, I. L. (eds), *Usability Evaluation in Industry*. London: Taylor & Francis.

WICKENS, C. D. (1992) *Engineering Psychology and Human Performance*. London: Harper Collins.

## Acknowledgement

Figures 1.8, 1.9, 1.10 and Table 1.1 are reproduced by kind permission of the British Standard Institution.

# PART ONE

# Methods in Product Design

# CHAPTER TWO

# Ergonomics methods in consumer product design and evaluation

NEVILLE STANTON and MARK YOUNG

*Department of Psychology, University of Southampton*

## 2.1 Introduction

There appears to be a growing number of texts in recent years describing, illustrating and espousing a plethora of ergonomics methods (Diaper, 1989; Kirwan and Ainsworth, 1992; Kirwan, 1994, Corlett and Clarke, 1995; Wilson and Corlett, 1995; Jordan *et al.*, 1996). This rise may be seen as a response to the requirement for more inventive approaches to assessing users and their requirements. In many ways this may be taken to mean that the call for user-centred design has been taken seriously by designers. However, this success has forced the ergonomics community to develop methods to assist the design of products and devices. This demand seems to have resulted in the pragmatic development of methods having priority over scientific rigour. In a recent review of ergonomics methods, Stanton and Young (1995) identified over sixty methods available to the ergonomist. The abundance of methods might be confusing for the ergonomist, Wilson (1995) goes as far as to suggest that a

> method which to one researcher or practitioner is an invaluable aid to all their work may to another be vague or insubstantial in concept, difficult to use and variable in its outcome. (p. 21)

This quote highlights the fact that most methods are used by their inventors only. Despite the proliferation of methods, there are few clues in the literature to enable ergonomists to identify which methods are appropriate for any given design activity. Given that most people tend to use their two or three favourite methods, independently of the problem that they are addressing (Stanton and Young, 1997; Baber and Mirza, this volume), the purpose of this chapter is to show substantive differences between methods. The aim of this chapter is to

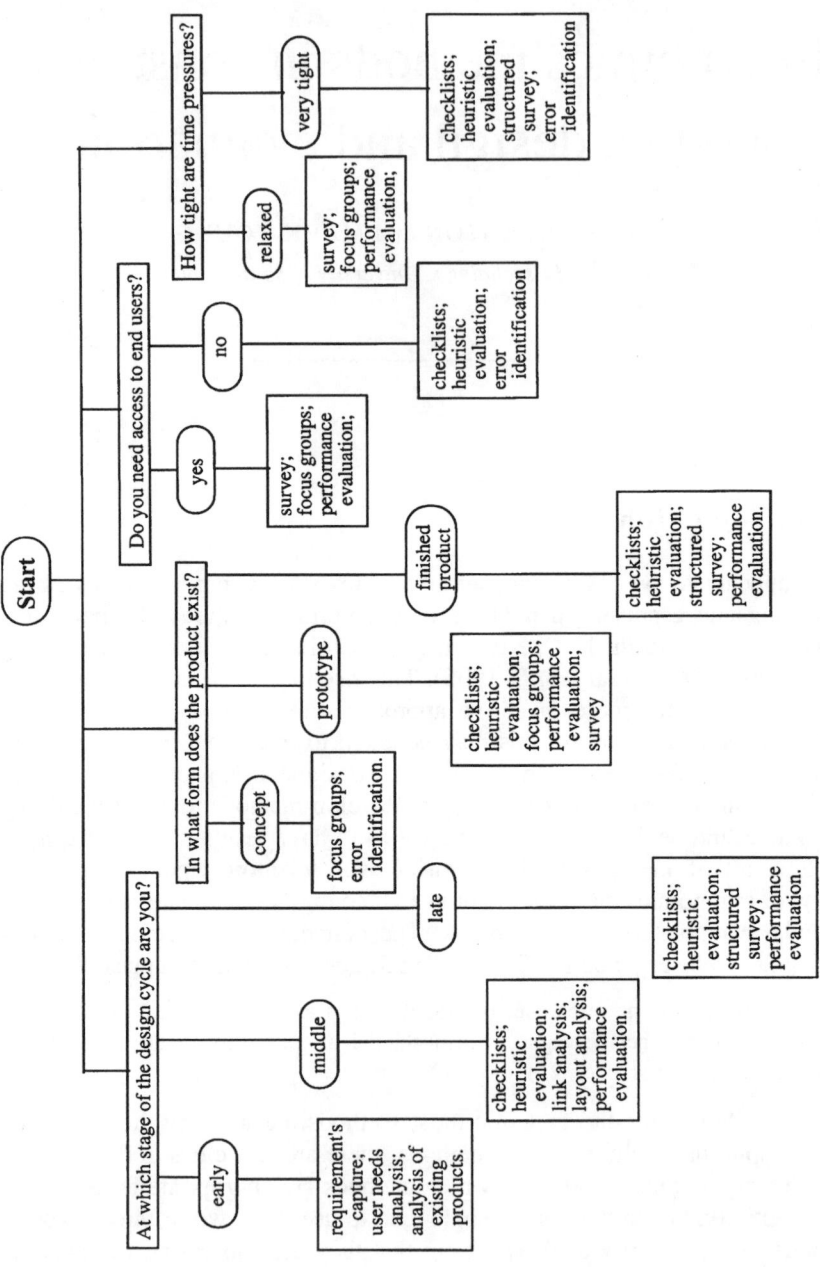

**Figure 2.1**  Some factors affecting the selection of methods (from Stanton and Baber, 1996a)

expose people to a variety of approaches to consumer product evaluation and to help give an informed judgement on their selection. Stanton and Baber (1996a) reduce the selection of methods down to four basic factors that it will depend upon:

- the stage of the design process;
- the form that the product takes;
- access to end-users;
- pressure of time (and other resources).

One or more of these factors will determine which methods are appropriate, as illustrated in Figure 2.1. From this analysis, Stanton and Baber (1996a) argue that it is not surprising that a checklist is the ubiquitous ergonomics method, as it appears to be the only method that is independent of all of these factors.

## 2.2  Review of Some Popular Methods

We have selected twelve methods for further consideration, based upon our analysis that these are a representative spread of methods that are currently being used to evaluate human/machine performance and assess the demands and effects upon people (Wilson, 1995). They were also chosen because of the appropriateness to the assessment of consumer products and user activity. Methods selected were:

- Heuristics
- Checklists/guidelines
- Observation
- Interviews
- Questionnaires
- Link analysis
- Layout analysis
- Hierarchical task analysis
- Systematic human error reduction and prediction approach (SHERPA)
- Task analysis for error identification
- Repertory grids
- Keystroke level model.

The aim of the review was to evaluate a device with each method in turn to determine the inputs (those materials and activities required to start the evaluation), processes (the activities of evaluation) and outputs (the resultant information supplied by the evaluation). Given the number of methods to be covered, emphasis will be given to providing an example of each approach with some accompanying text and reference to source material on the approach. The review is based upon the application of the methods to the evaluation of a car radio cassette player. Figure 2.2 indicates the main control display functions of the machine.

### 2.2.1 Heuristics

Heuristics require the analyst to use judgement, intuition and experience to guide the product evaluation (Nielsen, 1992). This method is wholly subjective and the output is likely to be extremely variable. In favour of the heuristic approach is the ease and speed with which it may be applied. Several techniques incorporate the heuristic approach (e.g. checklists, guidelines, SHERPA) but serve to structure heuristic judgement.

A heuristic analysis was applied to a typical car radio (Figure 2.3). Needless to say, the analysis was very quick to execute, and did not require any special knowledge on the analyst's part. Indeed, in many ways it resembled a simple walkthrough. Problems encountered in the analysis were few; the only real dissatisfaction residing in the fact that most of the output pertained to very similar faults with the device. Regarding the output, it may be observed that it is largely similar to the checklist approach in form and content, particularly as here we also see many items concerned with anthropometrics (e.g. button sizes). In addition, one of the remedial suggestions was also proposed in the SHERPA analysis (although this particular section of SHERPA is also

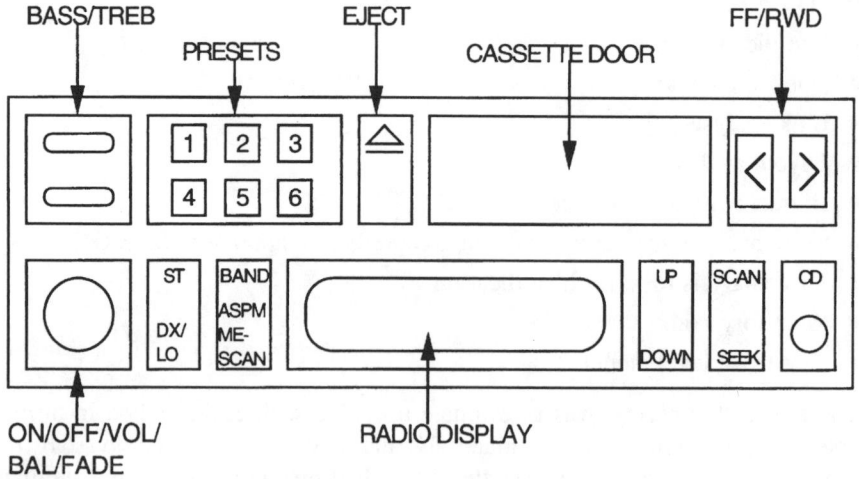

**Figure 2.2** A car radio cassette player.

- On/Off/Volume control is a tad small and awkward, combined with difficult Balance control
- Pushbutton operation would be more satisfactory for On/Off, as Volume stays at preferred level
- Fader Control is particularly small and awkward
- Both of the above points are related to the fact that a single button location has multiple functions - this is too complex
- Treble and Bass Controls also difficult and stiff; although these functions are rarely adjusted once set
- Station Preset Buttons are satisfactory; quite large and clear
- Band Selector Button and FM Mono-Stereo Button should not have 2 functions on each button - could result in confusion if wrong function occurs. These buttons are the only buttons on the radio which are not self-explanatory - the user must consult the manual to discover their function
- Tuning Seek and Tuning Scan Buttons are easier to understand and use, although there are still two functions on the same button. These are probably used more than the aforementioned buttons
- Cassette FF, RWD and Eject Buttons are self-explanatory; the same accepted style that is on all car radio designs. FF and RWD Buttons could be a little larger
- Auto-reverse function is not so obvious, although it is an accepted standard (pressing FF and RWD Buttons simultaneously)
- Illumination - is daytime/nighttime illumination satisfactory? A dimmer control would probably aid matters

**Figure 2.3** Heuristic analysis of the radio cassette.

subjective). Thus there is some overlap between these approaches, which can only be encouraging for heuristics.

## 2.2.2 Checklists/Guidelines

Checklists and guidelines would seem to be a useful *aide-mémoire*, to make sure that the full range of ergonomics issues have been considered (Ravden and Johnson, 1989; Woodson *et al.*, 1992). However, the approach may suffer from a problem of situational sensitivity, i.e. the discrimination of an appropriate item from an inappropriate one largely depends upon the expertise of the analyst. Nevertheless, checklists offer a quick and easy method for device evaluation.

Woodson's (1981) Transport Checklist turned out to be more of a set of guidelines, so could not be used as an example in this analysis. A cursory

The following are those items from the Human Engineering Design Checklist (Woodson, 1981) which are relevant and/or marginal (unsatisfactory) for the car radio under analysis.

**4. Console and panel design**
4.1 Displays
4.1.1 Principles
> e. Crucial visual checks identified by attention-getting devices (e.g. visual or aural signals).
> g. Probability of confusion among instruments is minimal.

4.1.2 Labeling
> a. Trade names and other irrelevant information deleted.
> b. Easy to read under expected conditions of illumination.

4.1.4 Scales, dials, counters
> a. Numbers and letters are large enough for accurate reading at normal distance.
> b. Reflected light does not create illusion warning is "ON" or obscure reading.

4.1.6 Indicator and legend lights
> k. Displays are arranged in relation to one another to reflect the sequence of use or the functional relations of the components they represent, in that order of preference.
> l. Distinct, functional areas set apart for purposes of ready identification are outlined by black lines...
> p. When transilluminated indicators are used under varied ambient illumination, a dimming control is provided.
> t. Provision is made for bulb removal from the front of the display panel without the use of tools, or other equally rapid and convenient means.
> w. Button surfaces are concave to fit the finger, or provide a high degree of frictional resistance to prevent slipping.
> x. Buttons provide "snap feel" or an audible click to indicate that the control has been activated.
> y. A channel or cover guard is provided when prevention of accidental activation is imperative.
> bb. Button displacement is between one-eighth inch and three-quarter inches.
> cc. Button resistance is between 10 ounces and 40 ounces.

4.1.8 Levers
> m. For knobs grasped by the fingertips:
> > 1) Minimum depth is at least three-quarter inch
> > 2) Minimum diameter is at least one inch
> q. Maximum resistance for fingertip operation of small (1-inch diameter) knobs does not exceed 4 and one-half ounces.

**Figure 2.4**  Checklist applied to the radio cassette.

---

4.2 Control/Display Relationships
4.2.1 Arrangements
   a. All controls having sequential relations, or having to do with a particular function or operation, or which are operated together, are grouped together, along with the associated displays.
   b. When a control is associated with a transilluminated indicator, the indicator is located so as to be immediately and unambiguously associated with the control.
   c. When a control is associated with a transilluminated indicator, the indicator is located above the control.
   d. If a control knob is adjacent to the instrument it controls, it is located so that the control or the hand normally used for setting does not obscure the indicator.
4.2.2 Precautions
   a. The control is located or oriented so that the operator is not likely to hit it or move it accidentally in the normal sequence of control movements.
   d. Interlocks are provided so that extra movement of the prior operation of a related or locking control is required.
   e. Resistance is built into the control so that definite or sustained effort is required to actuate it.

---

**Figure 2.4** *Continued.*

inspection of the vehicular related guidelines revealed that such recommendations are largely concerned with anthropometric improvements and safety (e.g. ingress and egress; protruding units, etc.). Thus these can only deal with issues such as consequentiality of accidents, comfort and satisfaction, rather than error prediction and usability (although comfort and satisfaction are elements of usability, there are also more cognitive components). The Human Engineering Design Checklist proved more useful. Only one section was deemed relevant to assessing a car radio; this was the section on console and panel design (Figure 2.4). Even so, it was clear that this was constructed for control room assessments, and many of the items were simply irrelevant. However, it would not be an arduous task to extract the relevant items and thus construct a checklist for assessing in-car devices. Furthermore, some items which were not relevant to a car radio (e.g, those concerned with CRTs) may be applicable to other devices (e.g. navigation aids). Again, though, the checklist was for the most part concerned with anthropometric issues.

### 2.2.3 Observation

Observation is perhaps the most obvious way of collecting information about a person's interaction with a device; watching and recording the interaction will undoubtedly inform the analyst of what occurred on the occasion observed

| Task | Errors Observed | F1 | F2 | T1 (s) mean (sd) | T2 (s) mean (sd) |
|---|---|---|---|---|---|
| 1 | | | | 4.64 (4.38) | 4.05 (3.02) |
| 2 | Didn't turn knob enough to adjust | 3 | 1 | 10.5 (7.67) | 6.14 (1.88) |
| | Pressed Seek | 1 | | | |
| 3 | Adjusted Treble | 1 | | 20.4 (13.3) | 10.1 (3.75) |
| | Adjusted Volume | 2 | | | |
| | Pressed On/Off | 1 | | | |
| 4 | | | | 15.7 (15.5) | 8.55 (4.50) |
| 5 | Adjusted Fade | 2 | 5 | 18.1 (9.61) | 11.9 (5.53) |
| | Adjusted Bass | 1 | | | |
| | Didn't attempt - forgot how | | 1 | | |
| 6 | Used Seek | 1 | | 7.14 (9.88) | 3.86 (1.73) |
| 7 | Pressed Preset and Seek together | 1 | 1 | 31.5 (24.7) | 23.6 (10.7) |
| | Held Seek button down | 2 | 1 | | |
| | Interrupted Seek by pressing Preset | | 1 | | |
| | Pressed preset | | 1 | | |
| | Used Manual tuning | | 1 | | |
| | Failed to store - didn't know how | 10 | | | |
| | Didn't hold preset long enough | 10 | 4 | | |
| | Pressed Seek to store | 1 | | | |
| 8 | Didn't know function | 4 | | 44.2 (23.8) | 29.5 (13.6) |
| | Used Seek | 15 | 3 | | |
| | Held Seek button down | 2 | | | |
| | Pressed Preset | 1 | | | |
| | Failed to store - didn't know how | 7 | | | |
| | Didn't hold preset long enough | 10 | 4 | | |
| | Hit 2 Presets and storage failed | | 1 | | |
| 9 | | | | 3.64 (1.87) | 3.18 (1.33) |
| 10 | Failed to stop FF/RWD | 1 | 2 | 36.6 (19.2) | 30.9 (16.1) |
| | Pressed Seek instead of autoreverse | 3 | | | |
| | Pressed wrong direction | 3 | 2 | | |
| | Turned tape over manually | 6 | | | |
| | Failed to Seek | 1 | 1 | | |
| 11 | Pressed twice | | 1 | 4.55 (2.77) | 3.91 (1.69) |
| 12 | | | | 2.86 (1.21) | 2.05 (0.576) |

**Task list:**

1. Switch On
2. Adjust Volume
3. Adjust Bass
4. Adjust Treble
5. Adjust Balance
6. Choose a new Preset station
7. Choose a new station using Seek and store it
8. Choose a new station using Manual search and store it
9. Insert cassette
10. Find next track on other side of cassette
11. Eject cassette
12. Switch Off

**Figure 2.5** Observation of user activity with the radio cassette.

(Drury, 1995; Kirwan and Ainsworth, 1992; Baber and Stanton, 1996a). Observation is also a deceptively simple method: one simply watches, participates in, and/or records the interaction. However, the quality of the observation will largely depend upon the method of recording and analysing the data. There are concerns about the intrusiveness of observation, the amount of effort required in analysing the data, the objectivity of the analysis and the comprehensiveness of the observational method. Despite these concerns, it is difficult to manage without some form of observational data, as most ergonomics methods rely upon it, e.g. hierarchical task analysis and link analysis.

The observational studies show errors and response times from 30 participants performing a range of tasks on two occasions (Figure 2.5). These data may seem to be highly credible in the eyes of designers, but observational studies are very resource-intensive and provide little output regarding cognitive mechanisms. They are applicable only late in the design process. The wide use of the observational technique suggests that it is both reliable and valid, as well as useful. Such considerations need to be addressed in a study of usability evaluations.

## 2.2.4 Interviews

Like observation, the interview has a high degree of ecological validity associated with it: if you want to find out what a person thinks of a device you simply ask them (Cook, 1988; Sinclair, 1995; Kirwan and Ainsworth, 1992). Interviewing has many forms, ranging from highly unstructured (free-form discussion) through focused (a situational interview), to highly structured (an oral questionnaire). For the purposes of device evaluation, a focused approach would seem most appropriate. The interview is good at addressing issues beyond direct interaction with devices, such as the adequacy of manuals and other forms of support. The strengths of the interview are the flexibility and thoroughness it offers. For the purposes of this review we undertook the interview within the Ravden and Johnson (1989) framework of eleven areas of usability. This served as an interview agenda to focus the interviewer and respondent of usability issues associated with the operations of the radio cassette (Figure 2.6).

The most striking aspect about the interview was its speed of administration – the whole process lasted around thirty minutes. As its structure was based on the sections of an HCI checklist, we can be quite confident that it thoroughly covered all aspects of device interaction. Admittedly, some aspects of the checklist were simply inapplicable to a car radio, but this just affirmed one advantage of the interview – its flexibility in adapting to changing scenarios. The output of the interview was on the whole unsurprising; much of it resembling the output of the heuristic analysis. Again, though, there are a number of advantages to the structured approach. Thorough coverage has been

**Interview results**

SECTION 1: VISUAL CLARITY
*Information displayed on the screen should be clear, well-organised,*
*unambiguous and easy to read.*

- There is a certain amount of visual clutter on the LCD; particularly
  with respect to preset station number
- Little or no discrimination between functions
- Writing (labelling) is small but readable
- Labelling is all upper case
- Ambiguous abbreviations (e.g., DX/LO; ASPM ME-SCAN)
- Gear lever can obscure vision to controls

SECTION 2: CONSISTENCY
*The way the system looks and works should be consistent at all times*

- Tuning buttons (especially Scan and Seek functions) present
  inconsistent labelling
- Moded functions create problems in knowing how to initiate the
  function (i.e., 'press' vs 'press and hold')

SECTION 3: COMPATIBILITY
*The way the system looks and works should be compatible with user*
*conventions and expectations*

- 'Scan' and 'Seek' buttons lack compatibility
- 4 functions on 'On/Off' switch makes it somewhat incompatible
- Programming preset stations may not be intuitive for a novice user,
  however is compatible with other systems
- Auto-reverse function could cause cognitive compatibility problems,
  particularly when involving FF/RWD functions

SECTION 4: INFORMATIVE FEEDBACK
*Users should be given clear, informative feedback on where they are in the*
*system, what actions they have taken, whether these actions have been*
*successful and what actions should be taken next*

- Tactile feedback is poor, particularly for the 'On/Off' switch
- Instrumental and operational feedback generally good, except in the
  case of programming a preset station, when operational feedback is
  poor

**Figure 2.6**  Interviewing users about the radio cassette.

SECTION 5: EXPLICITNESS
*The way the system works and is structured should be clear to the user*

- The novice user may not understand how to program stations without consulting the manual
- Resuming normal cassette playback after FF or RWD is not clear
- Similarly, initiating the auto-reverse function is not obvious

SECTION 6: APPROPRIATE FUNCTIONALITY
*The system should meet the needs and requirements of users when carrying out tasks*

- Rotating dial is not appropriate for front/rear fader control; maybe a joystick control would be more apt
- Prompts for task steps may be useful when programming stations
- Radio is muted when tuning - perhaps it would be possible to monitor the tuning process

SECTION 7: FLEXIBILITY AND CONTROL
*The interface should be sufficiently flexible in structure, in the way information is presented and in terms of what the user can do, to suit the needs and requirements of all users, and to allow them to feel in control of the system*

- Novice users may experience some difficulty
- Users with larger fingers may find controls fiddly
- Radio is inaudible whilst winding cassette - this is inflexible

SECTION 8: ERROR PREVENTION AND CORRECTION
*The system should be designed to minimise the possibility of user error, with inbuilt facilities for detecting and handling those which do occur; users should be able to check their inputs and to correct errors, or potential error situations before the input is processed*

- There is no 'undo' function for stored stations
- Separate functions would be better initiated from separate buttons (e.g., tuning up/down)
- Balance/Volume control is conducive to errors

**Figure 2.6**  *Continued.*

mentioned; in addition it was possible to relate the responses to psychological issues of design (e.g. cognitive compatibility) as a domain expert was present. It could be argued that the subjective responses, combined with professional wisdom, make this a very strong technique. In particular, one aspect emerged

SECTION 9: USER GUIDANCE AND SUPPORT
*Informative, easy-to-use and relevant guidance and support should be provided, both on the computer (via an on-line help facility) and in hard-copy document form, to help the user understand and use the system*

- Manual is not well structured (no contents page; installation instructions are mixed up with operations)
- Relevant manual sections are not easy to find, but this is alleviated somewhat by the manual being short
- Instructions in the manual are matched to the task

SECTION 10: SYSTEM USABILITY PROBLEMS

- Minor problems in understanding function of 2 or 3 buttons
- Finding information in the manual can be problematic
- Writing (labelling) on the radio is small
- Operation of 'Scan' button can be misunderstood
- Treble and bass controls are tiny

SECTION 11: GENERAL SYSTEM USABILITY

- Best aspect: This radio is *not* mode-dependent
- Worst aspects: Ambiguity in button labelling; Tactile feedback on 'Volume' control could be improved
- Confusing/difficult aspects: Only on initial use of FF and RWD when in auto-reverse mode
- Irritating aspects: Volume control causes static interference
- Common mistakes: Adjusting balance instead of volume
- Recommended changes: Substitute pushbutton operation for 'On/Off' control

**Figure 2.6** *Continued.*

from this analysis which had not been covered by any of the other techniques – the usability of the instruction manual.

### 2.2.5 Questionnaires

There are few examples of standardised questionnaires appropriate for the evaluation of in-car devices. However, the system usability scale (SUS) may, with some minor adaptation, be appropriate. SUS was developed as part of the usability engineering programme in integrated office systems developed at the Digital Equipment Company. SUS comprises ten items which relate to the usability of the device. Originally conceived as a measure of software usability, it has some evidence of proven success. The distinct advantage of this approach is the ease which the measure may be applied. It takes less than

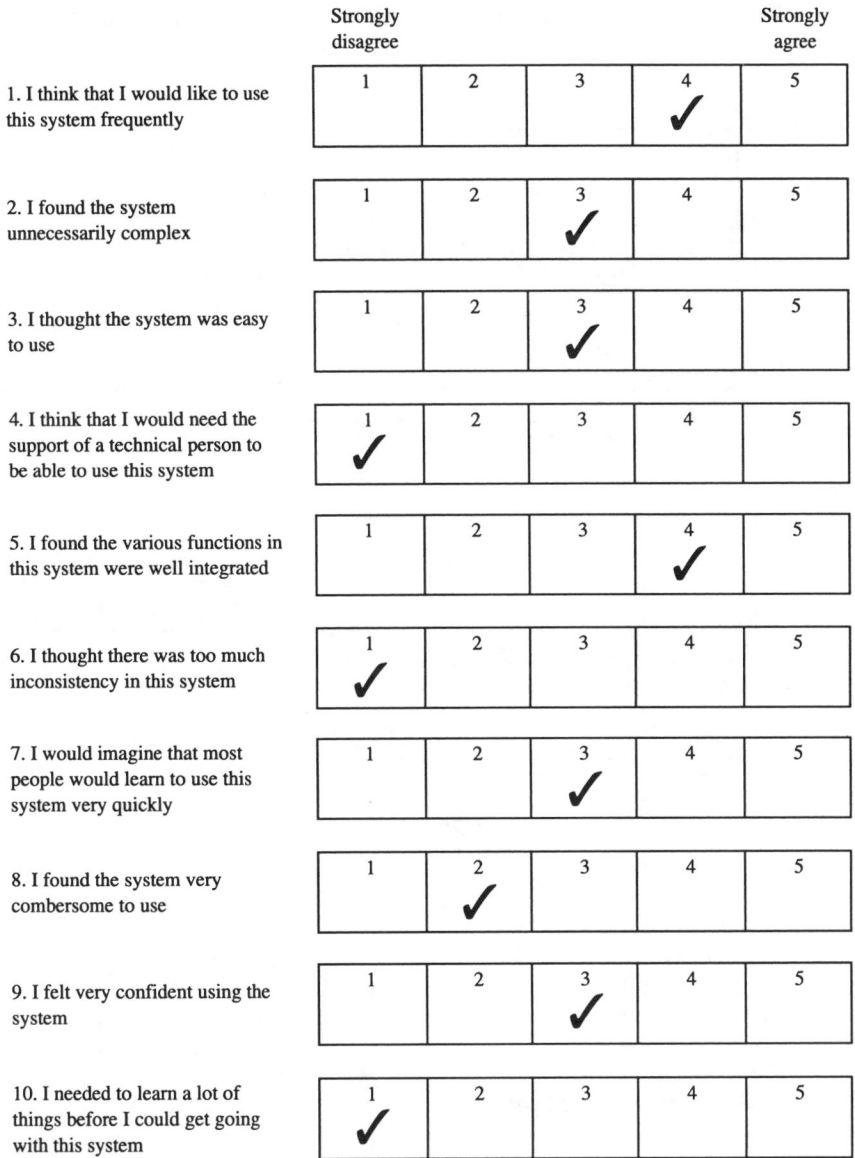

**Figure 2.7**   Rating the radio cassette using the system usability scale.

a minute to complete the questionnaire and no training is required (Brooke, 1996).

The SUS score is determined by subtracting 1 from all the scores on items with odd numbers and subtracting the scores from 5 for all the items with even numbers. The resultant value is multiplied by 2.5 to give an overall SUS rating of between 0 (extremely poor usability) and 100 (excellent usability). In the example shown in Figure 2.7 this resulted in the value 29 multiplied by 2.5,

Initial design:

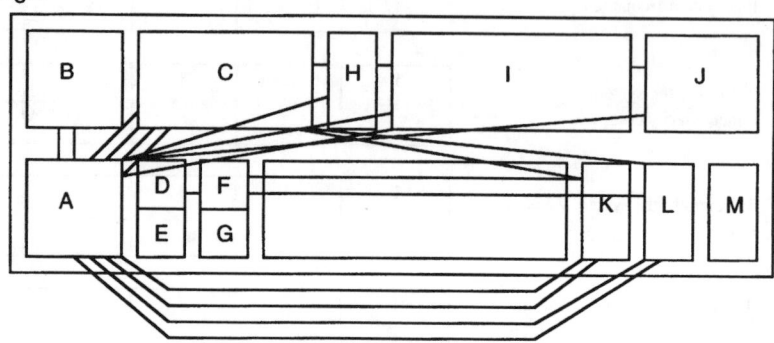

Revised design:

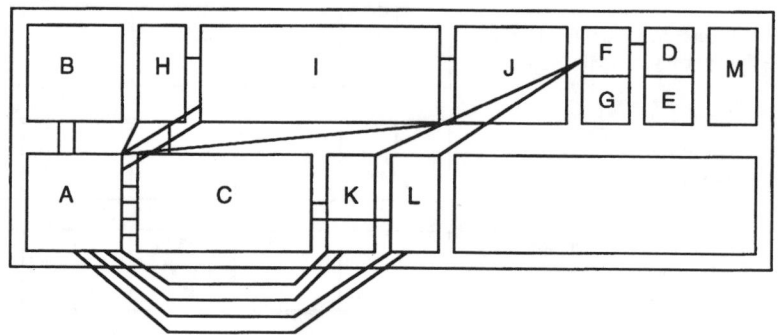

Key:

A = On-Off/Volume/Balance/Fader

B = Treble/Bass

C = Station Preset Buttons

D = FM Mono-Stereo Button

E = DX-Local Button

F = Band Selector Button

G = ASPM/Preset Memory Scan Button

H = Tape Eject Button

I = Cassette Compartment

J = Fast Wind/Programme Buttons

K = Tuning Up/Down Buttons

L = Tuning Scan/Seek Buttons

M = CD Input Socket

(a)

**Figure 2.8**   Link analysis.

**Link Table**

|   | B | C | D | E | F | G | H | I | J | K | L | M |
|---|---|---|---|---|---|---|---|---|---|---|---|---|
| A | 2 | 4 | - | - | - | - | 1 | 2 | 1 | 2 | 2 | - |
| B |   | - | - | - | - | - | - | - | - | - | - | - |
| C |   |   | - | - | - | - | 1 | - | - | 1 | 1 | - |
| D |   |   |   | - | 1 | - | - | - | - | - | - | - |
| E |   |   |   |   | - | - | - | - | - | - | - | - |
| F |   |   |   |   |   | - | - | - | - | 1 | 1 | - |
| G |   |   |   |   |   |   | - | - | - | - | - | - |
| H |   |   |   |   |   |   |   | 1 | - | - | - | - |
| I |   |   |   |   |   |   |   |   | 1 | - | - | - |
| J |   |   |   |   |   |   |   |   |   | - | - | - |
| K |   |   |   |   |   |   |   |   |   |   | - | - |
| L |   |   |   |   |   |   |   |   |   |   |   | - |

(b)

**Figure 2.8**   *Continued.*

giving a SUS rating of 72.5. On its own this offers only a subjective value, but the rating could be of most use when a dozen or more participants are rating two or more products. Statistical analysis of the results could be used to indicate real differences between the products.

Given the brevity of the approach, it is likely to serve as a useful adjunct to other methods. Brooke (1996) reports that SUS is a reliable measure and correlates well with other subjective measures.

### 2.2.6   Link Analysis

Link analysis represents the sequence in which device elements are used in a given task or scenario. The sequence provides the links between elements of the device interface (see Figure 2.8). This may be used to determine if the current relationship between device elements is optimal in terms of the task sequence. Time data recorded on duration of attentional gaze may also be recorded in order to determine if display elements are laid out in the most efficient manner. The link data may be used to evaluate a range of alternatives before the most appropriate arrangement is accepted (Stammers *et al.*, 1990; Kirwan and Ainsworth, 1992; Drury, 1995).

An initial dilemma was encountered in link analysis in whether to study hand or eye movements. For simplicity, the analysis was restricted to hand movements. A basic walkthrough was used for the data collection. One particular problem here was concerned with the fact that operating a radio is

Initial design:

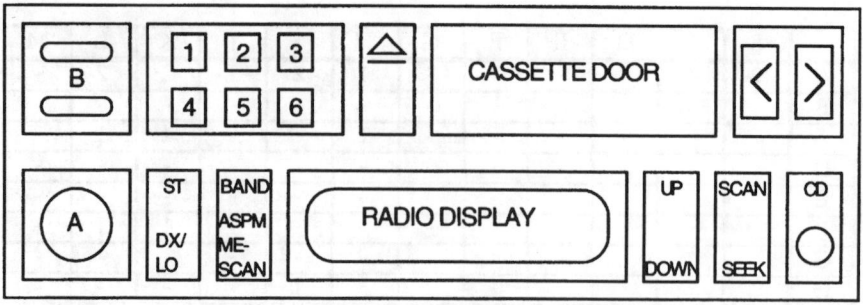

Functional groupings:

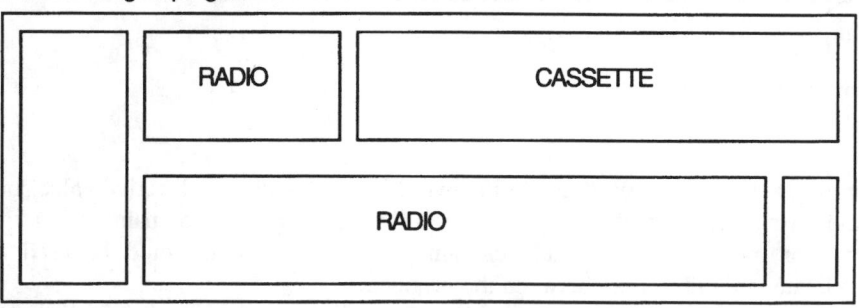

Importance of use:

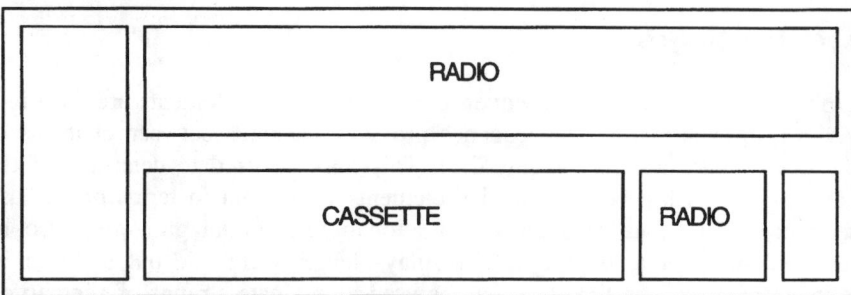

**Figure 2.9** A layout analysis.

Sequence of use (unchanged):

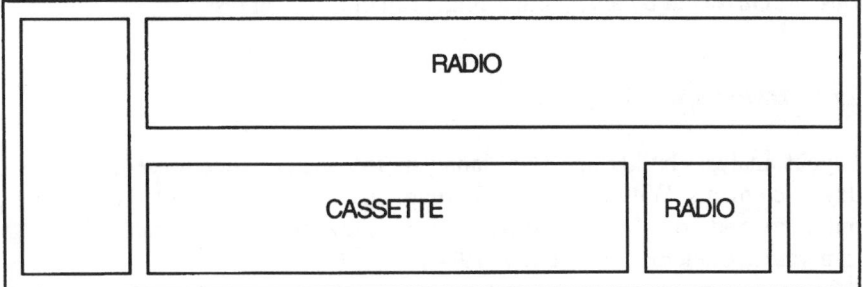

Within functional groupings:

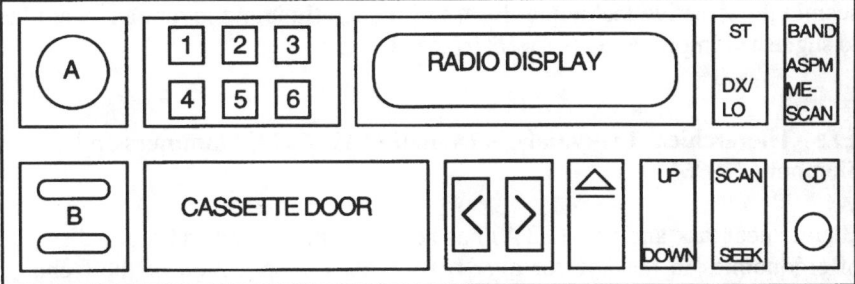

Revised design by importance, frequency and sequence of use:

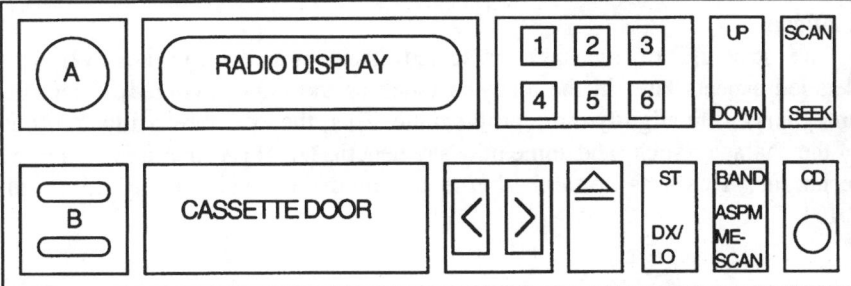

**Figure 2.9** *Continued*.

far from being a set procedure, leading to a possibly infinite set of links. This was circumvented by analysing a single run of a typical task.

### 2.2.7  Layout Analysis

Layout analysis builds upon link analysis to consider functional groupings of device elements. Within functional groupings, elements are sorted according to optimum trade-off of three criteria: frequency of use, sequence of use and importance of element (Easterby, 1984).

Layout analysis was undoubtedly easier to execute than link analysis. Functional groupings on a radio are obvious, and their importance, sequence and frequency of use were also easily determined. The analysis also maintains something of a hierarchical structure, as it progresses from general categories to specific functions (Figure 2.9). Overall, this was a very straightforward and seemingly effective technique. Both techniques (link and layout analysis) lead to suggested improvements for interface layout.

### 2.2.8  Hierarchical Task Analysis (Annett et al., 1971; Stammers and Shepherd, 1995)

Hierarchical task analysis (HTA) has been a technique central to the discipline of ergonomics in the UK for over two decades. Application of the technique breaks down tasks into goals, plans and operations in a hierarchical structure. Whilst the technique offers little more than a task description, it serves as the input to other predictive methods, for example SHERPA and keystroke level model. The concepts of HTA are relatively straightforward, but the approach requires some practice and reiteration before HTA can be applied with confidence.

The structure of the task presented itself immediately; however some detailed aspects later in the analysis (such as the logic involved in decisions and plans) were slightly more problematic. This, though, merely illustrates one of the characteristics (and some may say benefits) of HTA, that it is an iterative technique. This was certainly borne out in the analysis conducted (Figure 2.10).

### 2.2.9  Systematic Human Error Reduction and Prediction Approach

Systematic human error reduction and prediction approach (SHERPA) is a semi-structured human error identification technique. It is based upon hierarchical task analysis (HTA) and an error taxonomy. Briefly, each task step from the bottom-level in HTA is taken in turn and potential error modes

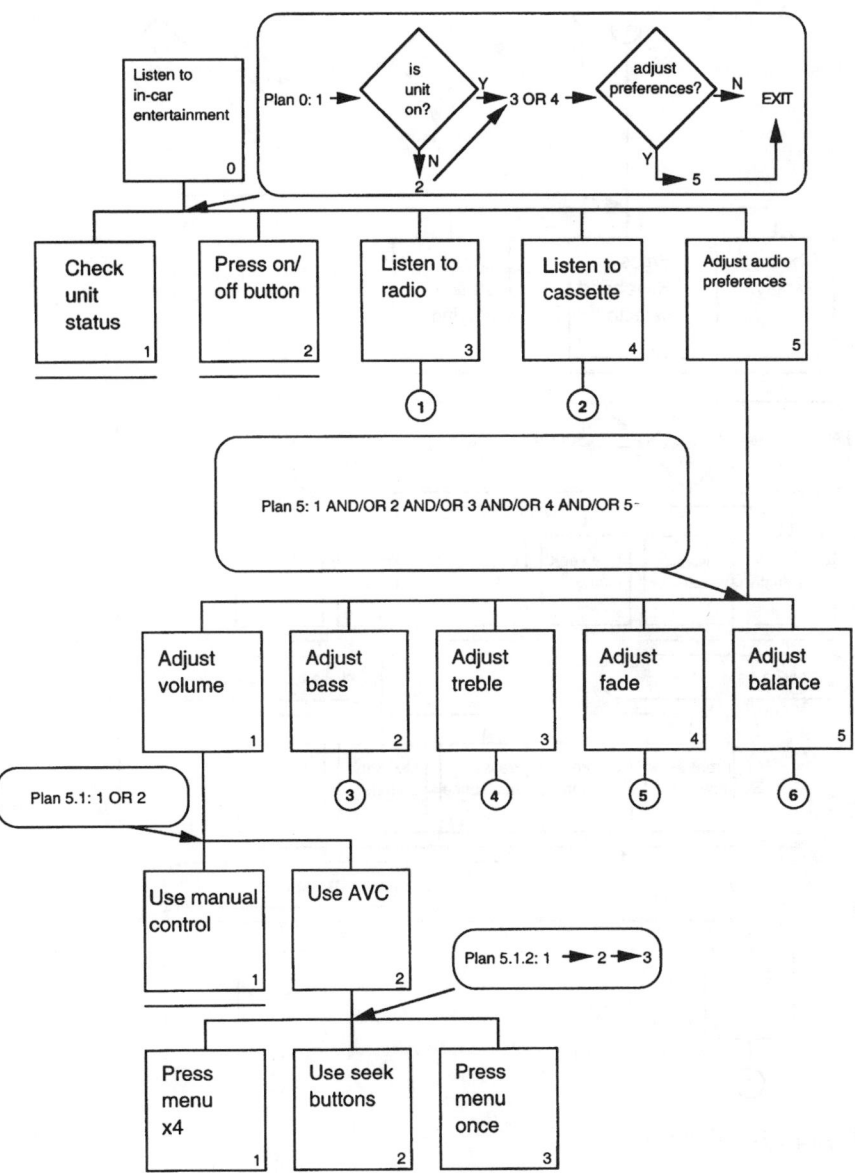

**Figure 2.10**  Hierarchical task analysis.

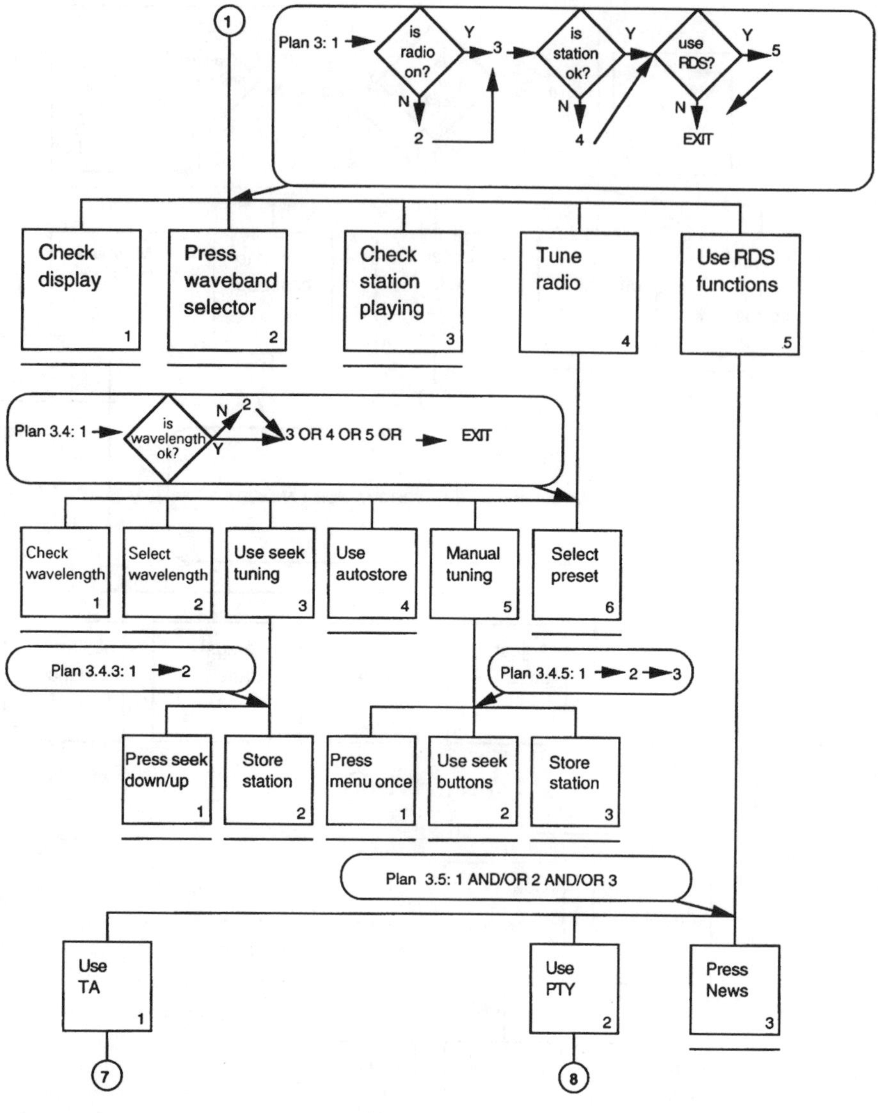

**Figure 2.10** *Continued.*

associated with that activity are identified. From this the consequences of those errors are determined. SHERPA appears to offer reasonable predictions of performance but may have some limitations in its comprehensives and generalisability (Embrey, 1983; Stanton, 1995; Baber and Stanton, 1996b).

Whilst the human error taxonomy incorporated in SHERPA is certainly a handy prompt in identifying errors, it does have limitations in generalisation

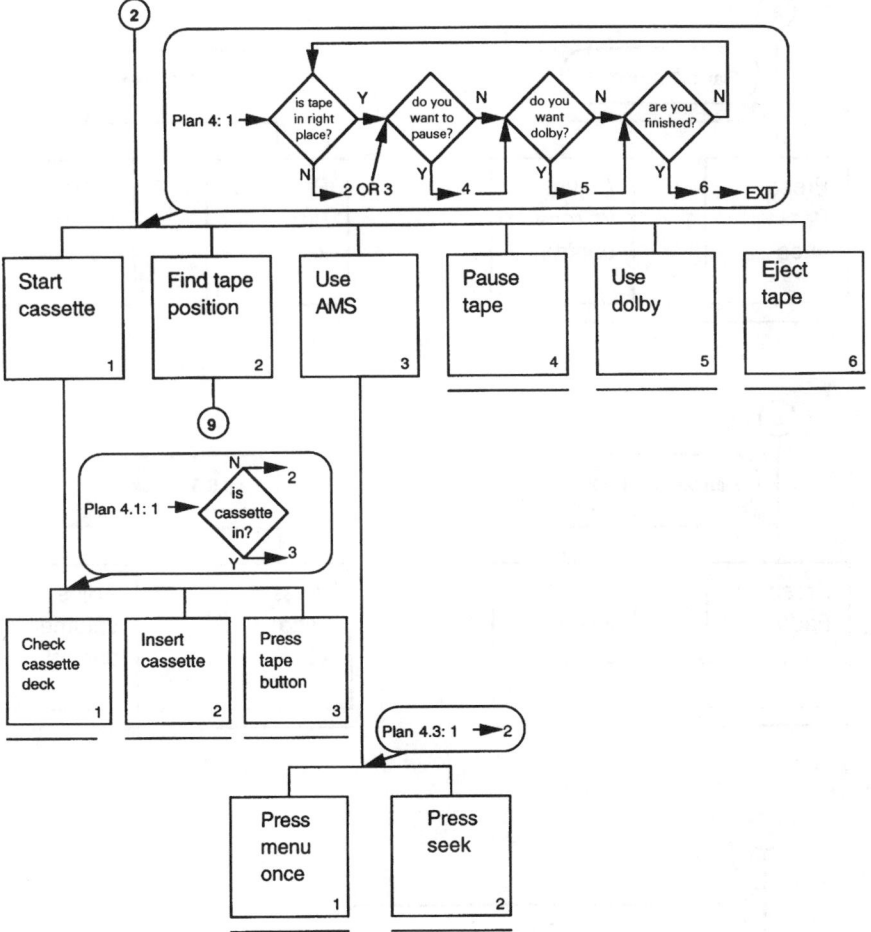

**Figure 2.10** *Continued.*

across tasks. As SHERPA was originally designed for process control room tasks, some sections of the taxonomy are quite inappropriate for product design and evaluation. Thus, as stated above, there is room for improvement in this area. The error predictions and remedies offered by SHERPA are certainly of applied value if proved to be valid. Indeed, there are some recent studies (e.g. Baber and Stanton, 1996b; Stanton and Stevenage, 1997) which suggest such predictions are to a large extent accurate and reliable. However, the credibility and salience of some of the predicted errors may be questioned. Even after a thorough SHERPA, error reduction strategies must be evaluated along dimensions such as practicability and cost-effectiveness to determine whether they are worth applying. Moreover, cognitive aspects of errors, which may aid

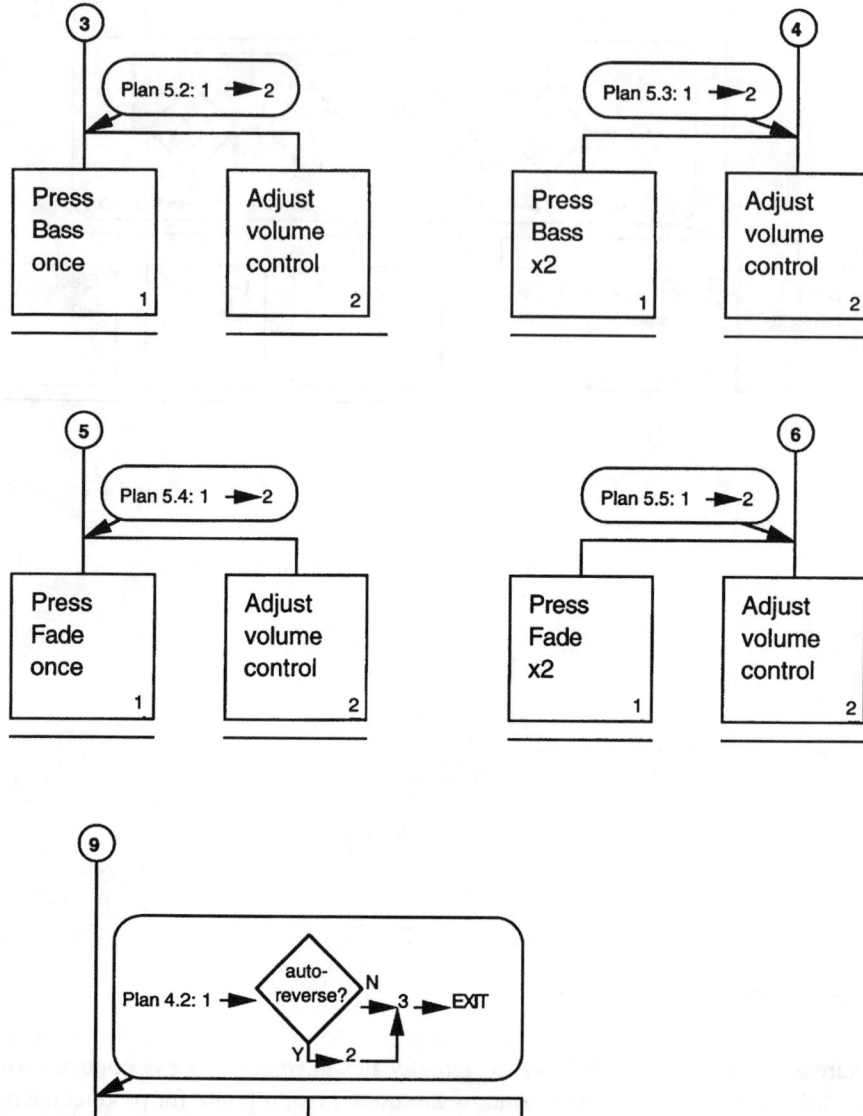

**Figure 2.10** *Continued.*

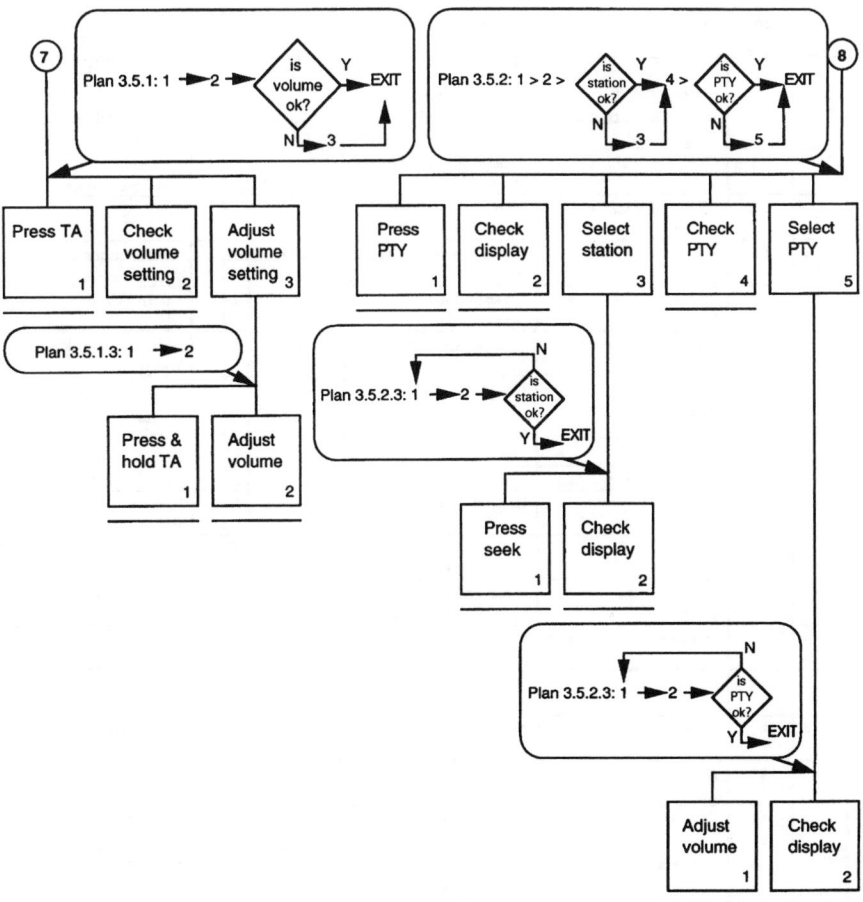

**Figure 2.10**   *Continued.*

in the above determination, are largely avoided by SHERPA (although differing error mechanisms were intimated when multiple error modes lead to similar consequences). Finally, an observation worthy of note is that executing the SHERPA served to highlight deficiencies in the HTA – undoubtedly a bonus for task analysis. The SHERPA results of the car radio cassette are given in Figure 2.11.

## 2.2.10   Task Analysis for Error Identification

Task analysis for error identification (TAFEI) is an approach for modelling the interaction between device and user. It is based upon hierarchical task analysis

| Task Step | Error Mode | Description | Consequence | Recovery | P | C | Remedies |
|---|---|---|---|---|---|---|---|
| 1 | A4 | Volume level adjusted inappropriately | Volume is at undesirable level | Immediate /4.2 | M | | Separate vol/on/off Preset startup volume |
| | A7 | Balance adjusted instead of on/off/vol | Unit is not switched on; balance settings altered | Immediate /4.3 | H | | Separate balance/on/off Lockout mechanism |
| 2.1 | C1 | Omit check of station | Listening to wrong station | 2.2 | L | | Untuned on startup Reminder to check |
| | C2 | Misidentify station | Listening to wrong station | 2.2 | M | | RDS |
| 2.2.1 | S2 | Select wrong preset | Listen to wrong station | 2.2.2 on | H | | Label buttons clearly Aide-mémoire |
| 2.2.2 | A1 | Preset button held too long | Undesired station storage | 2.3 | L | | Confirm before storage |
| | A6 | Press wrong button | Desired station not found | Immediate | H | | Label buttons clearly Aide-mémoire |
| 2.3.1 | C1 | Omit check of wavelength | Unnecessary retuning | 2.3.2 | L | | Display conspicuity (e.g., RDS) |
| | C3 | Check wrong display | Wavelength not identified | 2.3.2 | L | | Display conspicuity (e.g., RDS) |
| 2.3.2 | A2 | Adjust wavelength late | Unnecessary retuning | 2.3.4 on | L | | Better manual/training |
| | A4 | Over-pressing button | Desired setting missed | Immediate | L | | Responsive buttons Better manual/training |
| | A6 | Press wrong button | Undesired action; desired wavelength omitted | Immediate | L | | Label buttons clearly |
| | A8 | Fail to change wavelength | Undesired wavelength; unnecessary retuning | 2.3.4 on | L | | Better manual/training |
| 2.3.3 | C1 | Omit check of frequency | Unnecessary retuning | 2.3.4 | L | | Display conspicuity (e.g., RDS) |
| | C3 | Check wrong display | Frequency not identified | 2.3.4 | L | | Display conspicuity (e.g., RDS) |
| 2.3.4 .1 | A1 | Auto-search too long | Miss desired frequency | Immediate | M | | Better manual/training Alter search parameters |
| | A2 | Adjust freq. before wavelength | Tuning undesired wavelength | Immediate /2.3.5 | L | | Confirmation prompt Better manual/training |
| | A6 | Press wrong button | Undesired action; desired frequency omitted | Immediate | M | | Label buttons clearly Clarify functions |
| | A8 | Fail to adjust frequency | Undesired frequency; unnecessary retuning | 2.3.5 on | L | | Better manual/training |
| | A9 | Incomplete search | Undesired frequency; unnecessary retuning | 2.3.5 on | L | | Better manual/training |
| 2.3.4 .2 | A1 | Press button too long | Switches to auto-search | Immediate | H | | Alter button function Clarify function |
| | A2 | Adjust frequency before wavelength | Tuning undesired wavelength | Immediate | L | | Better manual/training Confirmation prompt |
| | A3 | Tuning in wrong direction | Tuning inefficiently | Immediate | M | | Increase compatibility Label buttons clearly Use knob instead |
| | A4 | Over-zealous use of button | Miss desired frequency | Immediate | M | | Increase buffer |
| | A6 | Press wrong button | Undesired action; desired frequency omitted | Immediate | M | | Label button clearly Use knob instead |
| | A7 | Use manual button for auto-search | Inefficient tuning | Immediate | M | | Label button clearly Use knob instead |
| | A8 | Fail to adjust frequency | Undesired frequency | 2.3.5 on | L | | Better manual/training |

**Figure 2.11** SHERPA.

| | | | | | | | |
|---|---|---|---|---|---|---|---|
| | A9 | Incomplete frequency search | Undesired frequency; unnecessary retuning | 2.3.5 on | L | | Better manual/training |
| | A10 | Auto-search on wrong button | Undesired action; desired omitted | Immediate | L | | Label buttons clearly Use tuning knob |
| 2.3.5 | A1 | Button press too short | Station not stored | Immediate | M | | Better manual/training Decrease threshold time |
| | A2 | Store station before freq | Inappropriate storage | 2.2 on | L | | Better manual/training |
| | A6 | Press wrong button | Station not stored/stored inappropriately | 2.2 on | M | | Label buttons clearly |
| | A7 | Change station instead of storing | Station not stored; frequency lost | 2.3 on | M | | Better manual/training Confirmation prompt |
| | A8 | Fail to store station | Station lost | 2.2 | L | | Confirmation prompt Auto-store prompt |
| | A10 | Change to wrong station & not store | Undesired station; frequency lost | 2.2 | L | | Confirmation prompt Auto-store prompt |
| 3.1 | A5 | Cassette insertion misaligned | Cassette will not fit | Immediate | L | | Label entry door more clearly |
| 3.2 | C1 | Omit check of cassette | Cassette in wrong position | 3.3 on | M | | Monitor/counter Clearer direction display |
| | C2 | Incomplete check on cassette | Cassette in wrong position | 3.3 on | M | | Monitor/counter Clearer direction display |
| 3.3 | A10 | Press FF/RWD buttons | Cassette in wrong position | 3.4 on | M | | Make autoreverse button distinct |
| 3.4 | A1 | FF too long | Miss desired point | 3.5 | H | | Monitor/counter |
| | A3 | RWD instead of FF | Miss desired point | 3.5 | L | | Label buttons clearly |
| | A6 | Press wrong button | Miss desired point | 3.5 | L | | Label buttons clearly |
| | A9 | Incomplete search | Miss desired point | 3.5 | L | | Monitor/counter |
| 3.5 | A1 | RWD too long | Miss desired point | 3.4 | H | | Monitor/counter |
| | A3 | FF instead of RWD | Miss desired point | 3.4 | L | | Label buttons clearly |
| | A6 | Press wrong button | Miss desired point | 3.4 | L | | Label buttons clearly |
| | A9 | Incomplete search | Miss desired point | 3.4 | L | | Monitor/counter |
| 4.1 | C1 | Omit check of audio settings | Audio settings undesired | 4.2 on | L | | Preset preferences Mute on startup |
| | C2 | Incomplete check of audio settings | Audio settings undesired | 4.2 on | L | | Preset preferences |
| 4.2 | A3 | Adjust volume wrong way | Undesired volume level | Immediate | L | | Compatibility Discrete functions |
| | A4 | Adjustment not enough/too much | Undesired volume level | Immediate | H | | Scaled volume levels Preset preferences |
| | A6 | Adjust balance/tone mistakenly | Undesired settings | 4.3 on | M | | Separate functions |
| | A7 | Turn off instead of down | Unit off unintentionally | Immediate | L | | Separate functions |
| 4.3 | A3 | Adjust balance wrong way | Undesired balance level | Immediate | L | | Compatibility Discrete functions |
| | A4 | Adjustment not enough/too much | Undesired balance level | Immediate | H | | Scaled balance levels Preset preferences |
| | A6 | Adjust volume/tone mistakenly | Undesired settings | 4.2 on | M | | Separate functions |
| 4.4 | A3 | Adjust tone wrong way | Undesired tone level | Immediate | L | | Compatibility Discrete functions |
| | A4 | Adjustment not enough/too much | Undesired tone level | Immediate | H | | Scaled tone levels Preset preferences |
| | A6 | Adjust volume/balance mistakenly | Undesired settings | 4.2 on | M | | Separate functions |

**Figure 2.11** *Continued.*

| 5 | A3 | Turn up instead of off | Volume too high; unit still on | Immediate | L | | Compatibility Separate functions |
|---|---|---|---|---|---|---|---|
| | A4 | Mute instead of off | Unit still on | None | M | | Eliminate mute Separate functions Increase conspicuousness |
| | A7 | Adjust volume instead of switching off | Undesired volume; unit still on | Immediate | L | | Separate functions |
| | A9 | Incomplete operation - switch not triggered | Unit still on | Immediate | L | | Increase conspicuousness Separate functions Improve switch |

**Figure 2.11**  *Continued.*

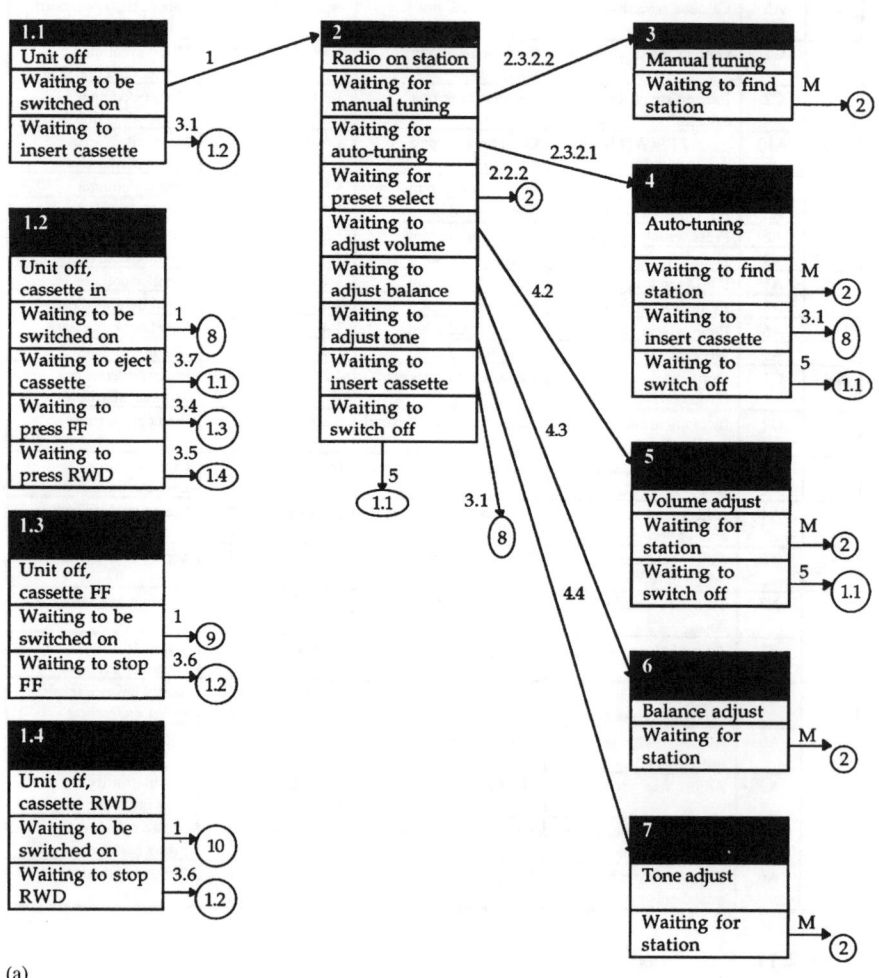

(a)

**Figure 2.12**  TAFEI.

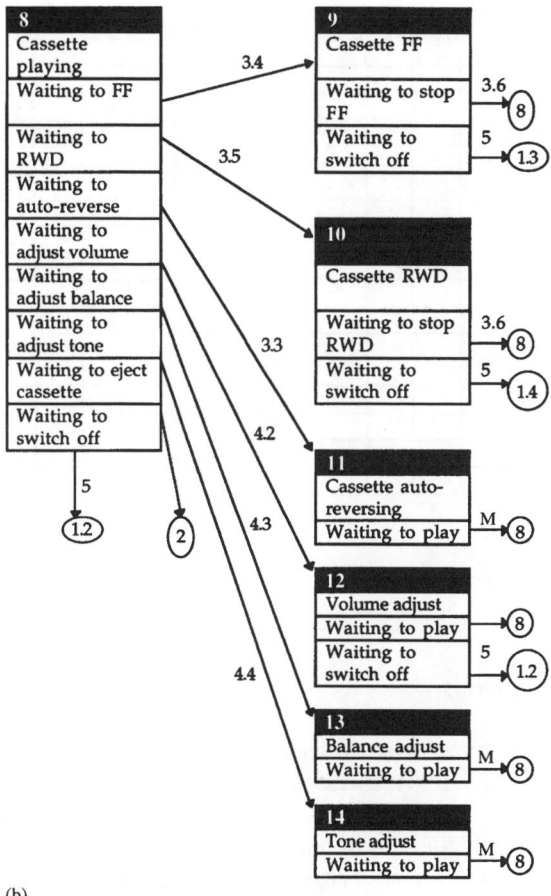

(b)

**Figure 2.12**   *Continued.*

(HTA) and state/space diagrams (SSDs), both established techniques with a pedigree of over 25 years. By mapping HTA onto SSDs and employing a transition matrix it is possible to start to consider what may go wrong in the interaction. In essence TAFEI is a human error identification method – like SHERPA (Baber and Stanton, 1994; Stanton and Baber, 1996b).

Once one has understood HTA and SSDs, executing a TAFEI analysis is not difficult, although it is still a little time-consuming (Figure 2.12). It is certainly an advantage to possess either or both HTA and SSD before beginning, as this saves a great deal of time. The most difficult part of this trial analysis was in constructing the SSDs for a car radio, as these have to be quite accurate for the remainder of the analysis to be effective. Completing the transition matrix is then a straightforward affair.

|      | 1.1 | 1.2 | 1.3 | 1.4 | 2 | 3 | 4 | 5 | 6 | 7 | 8 | 9 | 10 | 11 | 12 | 13 | 14 |
|------|-----|-----|-----|-----|---|---|---|---|---|---|---|---|----|----|----|----|----|
| 1.1  | -   | I   | -   | -   | L | - | - | - | - | - | - | - | -  | -  | -  | -  | -  |
| 1.2  | L   | -   | I   | I   | - | - | - | - | - | - | L | - | -  | -  | -  | -  | -  |
| 1.3  | -   | I   | -   | -   | - | - | - | - | - | - | - | L | -  | -  | -  | -  | -  |
| 1.4  | -   | I   | -   | -   | - | - | - | - | - | - | - | - | L  | -  | -  | -  | -  |
| 2    | L   | -   | -   | -   | - | L | L | L | L | L | L | - | -  | -  | -  | -  | -  |
| 3    | -   | -   | -   | -   | L | - | - | - | - | - | - | - | -  | -  | -  | -  | -  |
| 4    | L   | -   | -   | -   | L | - | - | - | - | - | L | - | -  | -  | -  | -  | -  |
| 5    | I   | -   | -   | -   | L | - | - | - | - | - | - | - | -  | -  | -  | -  | -  |
| 6    | -   | -   | -   | -   | L | - | - | - | - | - | - | - | -  | -  | -  | -  | -  |
| 7    | -   | -   | -   | -   | L | - | - | - | - | - | - | - | -  | -  | -  | -  | -  |
| 8    | -   | I   | -   | -   | L | - | - | - | - | - | - | L | L  | L  | L  | L  | L  |
| 9    | -   | -   | I   | -   | - | - | - | - | - | - | L | - | -  | -  | -  | -  | -  |
| 10   | -   | -   | -   | I   | - | - | - | - | - | - | L | - | -  | -  | -  | -  | -  |
| 11   | -   | -   | -   | -   | - | - | - | - | - | - | L | - | -  | -  | -  | -  | -  |
| 12   | -   | I   | -   | -   | - | - | - | - | - | - | L | - | -  | -  | -  | -  | -  |
| 13   | -   | -   | -   | -   | - | - | - | - | - | - | L | - | -  | -  | -  | -  | -  |
| 14   | -   | -   | -   | -   | - | - | - | - | - | - | L | - | -  | -  | -  | -  | -  |

(c)

**Figure 2.12**   *Continued.*

### 2.2.11   **Repertory Grids**

Repertory grids may be used to determine people's perception of a device (Kelly, 1955; Baber, 1996). In essence, the procedure requires the analyst to determine the elements (the forms of the product) and the constructs (the aspects of the product that are important to its operation). Each version of the product is then rated against each construct. This approach seems to offer a way of gaining insight into consumer perception of the device, but does not necessarily offer predictive information.

The repertory grid is not a difficult technique to execute, once the concept has been grasped (Figure 2.13). Constructing a thorough grid should take no more than an hour. However, analysis is a different story. Initially in this example, the revised analysis technique of Baber (1996) was attempted. Whilst it certainly seemed an easier method than the usual factor analysis, there also appeared to be weaknesses in its approach. These weaknesses were particularly borne out when it came to the factor extraction stage, for no constructs were significantly related to be grouped as a factor. Thus whilst the process of constructing the repertory grid was useful in itself, no useful quantification of the grid could be gleaned. Thus, the analysis turned to more conventional methods. It is quite possible to execute a number of analyses on a repertory grid, such as factor analysis, multidimensional scaling and even analyses of variance (although none of these methods was actually carried out in the present example). In summary, the repertory grid did provide a useful insight

| Constructs | Elements | | | | | | Opposites |
|---|---|---|---|---|---|---|---|
| | Rover | Ford | Vaux-hall | New Rover | Worst | Best | |
| Mode dependent | 1 | 5 | 4 | 1 | 5 | 1 | Separate functions |
| Pushbutton operation | 2 | 5 | 4 | 2 | 1 | 5 | Knob-turn operation |
| Unclear labelling | 2 | 5 | 4 | 2 | 5 | 1 | Clear labelling |
| Easy controls | 1 | 5 | 5 | 2 | 1 | 5 | Fiddly controls |
| Poor/no functional grouping | 4 | 5 | 2 | 2 | 5 | 1 | Good functional grouping |
| Good illumination | 2 | 4 | 5 | 2 | 1 | 5 | Poor illumination |

5 = left side very much applicable (right side not applicable at all)
4 = left side somewhat applicable (right side not really applicable)
3 = in between
2 = left side not really applicable (right side somewhat applicable)
1 = left side not applicable at all (right side very much applicable)
0 = characteristic irrelevant

**Figure 2.13**   Repertory grids.

into perception of this product, such that the output could be useful in design. However, the constructs elicited also seemed to mirror the output of some of the other techniques reviewed so far, such as checklists and heuristics.

## 2.2.12   Keystroke Level Model

The keystroke level model (KLM) is a technique that is used to predict task performance time for error-free operation of a device (Card *et al.*, 1983). The technique works by breaking tasks down into component activities, e.g. mental operations, motor operations and device operations, then determining response times for each of these operations and summing them. The resultant value is the estimated performance time for the whole operation. Whilst there are some obvious limitations to this approach (such as the analysis of cognitive operations) and some ambiguity in determining the number of mental operations to be included in the equation, the approach does appear to have some support. There are four motor operators in KLM – keystroking, pointing, homing and drawing; one mental operator; and one operator for system response. Each of these operators has an associated nominal time, derived by experiment (although drawing and response times are variable). It is thus a simple matter of determining the components of the task in question and summing the times of the associated operators to arrive at an overall task time prediction.

Although KLM is indeed a simplistic method, it was designed for HCI, and this is evident when attempting to apply it. With a car radio, an immediate

| Task | Operators | Associated times (s) | Total task time (s) |
|---|---|---|---|
| Select preset | M H K R | $1.35+0.4+0.2+Tr_0$ | 1.95 |
| Store station | M 2H 3K 2R | $1.35+(2*0.4)+$ $(3*0.2)+Tr_1+Tr_2$ | 6.25 |
| Operate cassette | M H 2K R | $1.35+0.4+(2*0.2)+$ $Tr_3$ | 7.15 |

Estimated System Response Times:

$Tr_0$ = negligible
$Tr_1$ = 1.5s
$Tr_2$ = 2.0s
$Tr_3$ = 5.0s

**Figure 2.14**   Keystroke level model.

stumbling block was encountered for there is no operator accounting for 'turning a knob'. Thus, whilst some operators have no place outside HCI (e.g. pointing, drawing), there are others which are not foreseen within HCI. A consequence of this is that we are immediately limited as to the tasks we can analyse in the automobile (Figure 2.14). Further restrictions are imposed regarding the fixed times associated with most operators, leading to some errant predictions – a time of 2 seconds simply for selecting a preset station seems generous. A final limitation of KLM is that it does not predict anything over and above task execution time, nor does it claim to.

## 2.3   Summary of Approaches

As can be seen, there is an immense variety in the range of approaches and in what they address (i.e. the human element, the device element or the interaction) and what they produce (e.g. task descriptions, predicted errors, performance times). For the purposes of considering where, in the design lifecycle, each of the methods was most appropriate, we summarised design to six main phases:

- *Concept* In which the idea for the device is considered in a largely informal manner, many implementations are considered and many degrees of freedom remain.
- *Flowsheeting* In which the ideas for the device become formalised and the alternatives considered become very limited.
- *Design* In which the design solution becomes crystallised and blueprints are devised.

**Table 2.1**  Methods in design stages

| Method | Design stage |
| --- | --- |
| Heuristics | All |
| Checklists | Design to operation |
| Observation | Prototype to operation |
| Interviews | All |
| Questionnaires | All |
| Link analysis | Prototype |
| Layout analysis | Prototype |
| HTA | Design to operation |
| SHERPA | Design to operation |
| TAFEI | Concept to design |
| Repertory grids | All |
| KLM | Concept to prototype |

- *Prototyping*  In which a hard built prototype device is developed for evaluation.

- *Commissioning*  In which the final design solution is implemented and enters the marketplace.

- *Operation and maintenance*  When the device is supported in the marketplace.

We present Table 2.1 as a guide to the applicability of each of the methods. This guide is only meant as a heuristic, and final choice will depend upon the circumstances surrounding the users and product under evaluation.

## 2.4  Conclusions

There is clearly little reported evidence in the literature of reliability or validity of ergonomics methods. The patterns of usage suggest that there is no clear match of methods to applications, which presents a rather confusing picture when embarking upon an evaluation. Apart from a few clearly defined applications, the pattern looks almost random. The detailed review of ergonomics methods led to a greater insight into the demands and outputs of the methods under scrutiny. A study by Stanton and Young (1997) indicated that link analysis, layout analysis, repertory grids and KLM appear to offer good utility when compared with other, more commonly used methods. We would suggest ergonomists and designers would be well served by exploring the utility of other methods rather than always relying upon three or four favourites. However, it is an important goal of future research to establish the reliability and validity of ergonomic methods. These data could provide the encouragement for designers to try alternative approaches.

## Acknowledgement

The research reported in this chapter was supported by the EPSRC under the LINK Transport Infrastructure and Operations Programme.

## References

ANNETT, J., DUNCAN, K. D., STAMMERS, R. and GREY, M. J. (1971) *Task Analysis*. Department of Employment Training Information Paper 6. London: HMSO.

BABER, C. (1996) Repertory grid theory and its application to product evaluation. In Jordan, P. W., Thomas, B., Weerdmeester, B. A. and McClelland, I. L. (eds), *Usability Evaluation in Industry*. London: Taylor & Francis, pp. 157–65.

BABER, C. and STANTON, N. A. (1994) Task analysis for error identification: a methodology for designing error-tolerant consumer products. *Ergonomics*, **37** (11), 1923–41.

BABER, C. and STANTON, N. A. (1996a) Observation as a usability method. In Jordan, P. W., Thomas, B., Weerdmeester, B. A. and McClelland, I. L. (eds), *Usability Evaluation in Industry*. London: Taylor & Francis, pp. 85–94.

BABER, C. and STANTON, N. A. (1996b) Human error identification techniques applied to public technology: predictions compared with observed use. *Applied Ergonomics*, **27** (2), 119–31.

BROOKE, J. (1996) SUS: a 'quick and dirty' usability scale. In Jordan, P. W., Thomas, B., Weerdmeester, B. A. and McClelland, I. L. (eds), *Usability Evaluation in Industry*. London: Taylor & Francis, pp 189–94.

CARD, S. K., MORAN, T. P. and NEWELL, A. (1983) *The Psychology of Human–Computer Interaction*. Hillsdale, NJ: Erlbaum.

COOK, M. (1988) *Personnel Selection and Productivity*. Chichester: John Wiley.

CORLETT, E. N. and CLARKE, T. S. (1995) *The Ergonomics of Workspaces and Machines* (2nd edition). London: Taylor & Francis.

DIAPER, D. (1989) *Task Analysis in Human Computer Interaction*. Chichester: Ellis Horwood.

DRURY, C. G. (1995) Methods for direct observation of performance. In Wilson, J. and Corlett, N. (eds), *Evaluation of Human Work* (2nd edition). London: Taylor & Francis, pp. 45–68.

EASTERBY, R. (1984) Tasks, processes and display design. In Easterby, R. and Zwaga, H. (eds), *Information Design*. London: Taylor & Francis.

EMBREY, D. (1983) Quantitative and qualitative prediction of human error in safety assessments, *Institution of Chemical Engineers Symposium Series*, **130**, pp. 329–50.

JORDAN, P. W., THOMAS, B., WEERDMEESTER, B. A, and MCCLELLAND, I. L. (1996) *Usability Evaluation in Industry*. London: Taylor & Francis.

KELLY, G. A. (1955) *The Psychology of Personal Constructs*. New York: Norton.

KIRWAN, B. and AINSWORTH, L. (1992) *A Guide to Task Analysis*. London: Taylor & Francis.

KIRWAN, B. (1994) *A Guide to Practical Human Reliability Assessment*. London: Taylor & Francis.

NIELSEN, J. (1992) Finding usability problems through heuristic evaluation. In *Proceedings of the ACM Conference on Human Factors in Computing Systems*. Monterey, CA: ACM Press, pp. 373–80.

RAVDEN, S. J. and JOHNSON, G. I. (1989) *Evaluating Usability of Human–Computer Interfaces: A Practical Method*. Chichester: Ellis Horwood.

SINCLAIR, M. (1995) Subjective assessment. In Wilson, J. and Corlett, N. (eds), *Evaluation of Human Work* (2nd edition). London: Taylor & Francis, pp. 69–100.

STAMMERS, R. B., CAREY, M. and ASTLEY, J. A. (1990) Task analysis. In Wilson, J. and Corlett. N. (eds), *Evaluation of Human Work*. London: Taylor & Francis, pp. 134–60.

STAMMERS, R. B. and SHEPHERD, A. (1995) Task analysis. In Wilson, J. and Corlett, N. (eds), *Evaluation of Human Work* (2nd edition). London: Taylor & Francis, pp. 144–68.

STANTON, N. A. (1995) *Analysing Worker Activity: A New Approach to Risk Assessment*, Health and Safety Bulletin, 240, pp. 9–11.

STANTON, N. A. and BABER, C. (1996a) Factors affecting the selection of methods and techniques prior to conducting a usability evaluation. In Jordan, P. W., Thomas, B., Weerdmeester, B. A. and McClelland, I. L. (eds), *Usability Evaluation in Industry*. London: Taylor & Francis, pp. 39–48.

(1996b) A systems approach to human error identification. *Safety Science*, 22 (1–3), 215–28.

STANTON, N. A. and STEVENAGE, S. (1997) Learning to predict human error: issues of reliability, validity and acceptability. *Ergonomics*, in press.

STANTON, N. A. and YOUNG, M. (1995) *Development of a Methodology for Improving Safety in the Operation of In-car Devices*, EPSRC/DOT LINK Report 1, University of Southampton.

(1997) Is utility in the mind of the beholder? A study of ergonomics methods. *Applied Ergonomics*. In press.

WILSON, J. (1995) A framework and context for ergonomics methodology. In Wilson, J. and Corlett, N. (eds), *Evaluation of Human Work* (2nd edition). London: Taylor & Francis, pp. 1–39.

WILSON, J. and CORLETT, N. (1995) *Evaluation of Human Work* (2nd edition). London: Taylor & Francis.

WOODSON, W. E., TILLMAN, B. and TILLMAN, P. (1992) *Human Factors Design Handbook* (2nd edition). New York: McGraw-Hill.

# CHAPTER THREE

# Evaluation of product safety using the BeSafe method

RACHEL BENEDYK and SARAH MINISTER

*Ergonomics and HCI Unit, University College London*

## 3.1 The Legal Requirement for Evaluation of Product Safety

The general understanding of a 'product' is a manufactured item, useful to a person for a purpose. Because products are directly used by people, any hazards or risks associated with the product usage would give rise to concern for the safety of users. There is therefore a requirement to minimise the hazards and risks of product use, which is embodied in various UK regulations such as the General Product Safety Regulations (GPSR), 1994 (Department of Trade and Industry, 1994) and the Provision and Use of Work Equipment Regulations, 1992 (Health and Safety Executive, 1992). There is similar concern whether the product is provided for use by employees or is sold to a consumer for domestic use; whether the product will be used by one individual or several; and whether the product is to be used in a fixed position or moved around. Producers and suppliers have responsibility to provide a product that must, as far as possible, be safe.

This implies a requirement to assess safety and safe usage of products. For example, the GPSR 1994 lay down a framework for assessing safety: 'Both producers and distributors have responsibilities to supply only products which are safe and to undertake relevant activities, including monitoring, where appropriate, to help ensure that a product remains safe throughout its reasonably foreseeable period of use.' In addition, 'a producer is also required... to adopt measures... to enable him to be informed of the risks which these products might present.' The GPSR 1994 define a safe product as 'any product which under normal or reasonably foreseeable conditions of use, including duration, presents no risk or only the minimal risk compatible with the product's use'. Reasonably foreseeable use should take account of the intended and potential types of use and 'how a reasonable person might use a product in the absence of any indications to the contrary'. Thus, potential abuse, or misuse, must be considered, whether intentional or not (Van Weperen, 1993).

Assessing the safety of a product (examples of typical methods are given in Stoop, 1987, and Schoone-Harmsen, 1990) must include, for example, its characteristics, its use under normal and occasional conditions (e.g. assembly), the effectiveness of the instructions provided for the consumer, the type of consumer put at risk and the views of the consumers themselves. Thus we see that there is considerable overlap between such a safety assessment and an ergonomic evaluation of product use (for examples of which see McClelland, 1995). An ergonomic approach – one centred on the user in the use environment – will therefore be beneficial when evaluating the safety of products.

Such regulations have been introduced in order to avoid the production of unsafe products. The probable consequence of an unsafe product is the presentation of risk to the user, and the ultimate consequence of this is an accident.

## 3.2  Formation of Accidents

Accidents are often caused as a direct result of risky behaviour by the user (Ryan, 1987a). If this risky behaviour is unintentional, it is being forced by other factors such as poor design or poor instructions. Moreover, users of products rarely recognise the hazards associated with product use (Ryan, 1987b). If the risky behaviour is intentional, it may be benign or malevolent. Benign intentional risks are usually driven by a need to short-cut procedures or to cope with too many demands at once. Malevolent risks are only taken by saboteurs and masochists.

We can see therefore that the aetiology of accidents can be complex, and the requirement of the regulations to reduce the risk of a product causing an accident is a challenging one. Some guidance can be provided by a systems approach, such as Reason's (1995) model of accident causation (Figure 3.1), created with reference to organisations.

In this ergonomic model, we see the 'blame' being taken off the person at the sharp end who made the error that caused the accident; instead the causes are multiple and shared throughout the organisation. Systems are not just operated by human beings, they are also designed, constructed, assembled, installed, organised, managed, maintained and regulated by them. Some potential causes for accidents lie remote in time or space from the event, and are the responsibility (sometimes unknowingly) of others in the organisation.

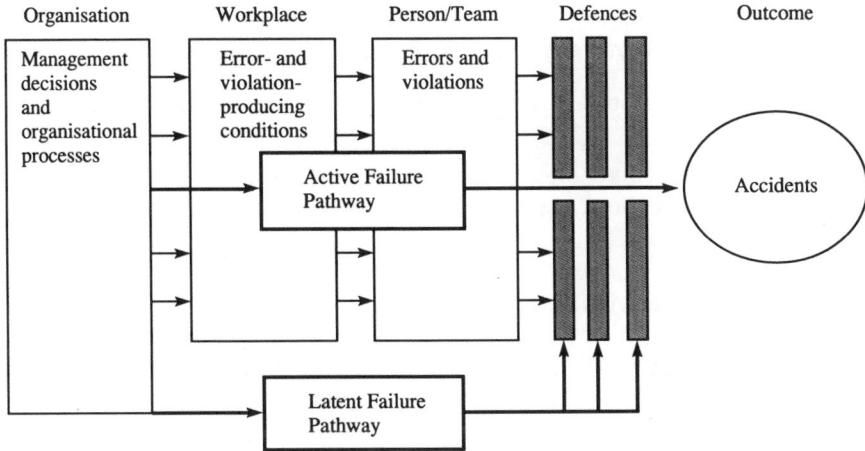

**Figure 3.1**   Reason's model of accident causation in organisations (Reason, 1995).

Thus, the inclination of some major accident investigations (e.g. Zagreb, Chernobyl, Three Mile Island, Kegworth) to pin the cause of the accident on operator error is an incomplete approach (for a discussion, see Pheasant, 1991). Prevention of similar accidents will require attention to more than just operator training or discipline.

Reason (1995) suggests that defences against human error by the operator are set up as part of the organisation's safety culture. Examples of such defences might be standards, procedures, controls or safety design features. Breaching these defences, either by error or violation, directly causes the accident: this is the *active failure*. However, predisposition towards the error or violation already exists within the workplace as a result of management decisions higher up the line. For example, if high workload causes the operator to make an error owing to fatigue, the cause lies with the setting of the workload. The organisational defence against such an error may be well in place (e.g. there are automatic braking systems that can stop a train whose driver has missed a red stop signal) or the defences may be weak, and an accident consequently happens. Weak defences are caused when the management level of the organisation makes poor decisions concerning, for example, forecasting, designing or maintaining: these are referred to as the *latent failures*. Active failures tell us what happened; latent failures tell us how and why.

Following an accident, an organisation can tackle active failures directly by means of warning or personal discipline to the worker or workers concerned, or extra training or supervision, but there can be no guarantee that there will not be another accident in the future. The potential for human error still remains. In the case of products, especially consumer products, even these measures against active failures may not be feasible. Thus latent failures are crucially important to accident prevention for two reasons (Simpson, 1994):

- If they are not resolved, the probability of repeat (or similar) accidents remains high.

- As one latent failure often influences several potential errors, removing latent failures can be a very cost-effective route to accident prevention.

## 3.3   The BeSafe Method for Safety Evaluation

The BeSafe ('*Be*havioral *Safe*ty') system was developed by ergonomists at British Coal as a means of targeting accident prevention initiatives (Simpson, 1994). Simpson suggested that Health and Safety Executive statistics show that 90 per cent of accidents at work are caused in some part by human error. Thus it is to this human behaviour that BeSafe is systematically addressed. BeSafe uses Reason's active and latent failure model as part of its basic framework.

BeSafe has been used in industry with a good level of success. In one coalmine where it was carried out, 48 potential or actual human errors were identified, categorised and acted upon. This led to an 80 per cent reduction in the accident rate (Simpson, 1994).

BeSafe (see Figure 3.2) is, in essence, an integrated set of ergonomics-based procedures (analysis techniques, checklists and questionnaires) which are designed to enable an auditor to identify systematically the potential for human error in a specified job, operation or system (Ergonomics and Safety Management Unit, 1995; Simpson *et al.*, 1997). The procedures firstly allow the identification of potential human errors likely to cause accidents (active failures). These might be directly caused by the design of the task in hand, or indirectly caused by the worker violating a procedure. The BeSafe procedures then enable the identification of organisational factors which are likely to predispose active failures; following analysis, these give the latent failures in the system. Once both potential errors and latent failings are known, the workers and management can together devise action plans for the reduction of human error in the system, addressing all factors that have been found likely to increase the probability of error.

Figure 3.2 gives an overview of the BeSafe process, together with an indication of the techniques used at various stages. In general, the developers' intention is that BeSafe techniques are designed for use by non-ergonomists following a short training course, although it may be useful to involve an ergonomics expert when devising action plans to reduce the human error potential. The method stresses the necessary involvement of the people doing, supervising, managing and designing the work, both in filling in the detail of the scenario at the start, and in devising plans to combat the potential problems at the end. It thus reflects Reason's perspective that human error links to all levels in the organisation.

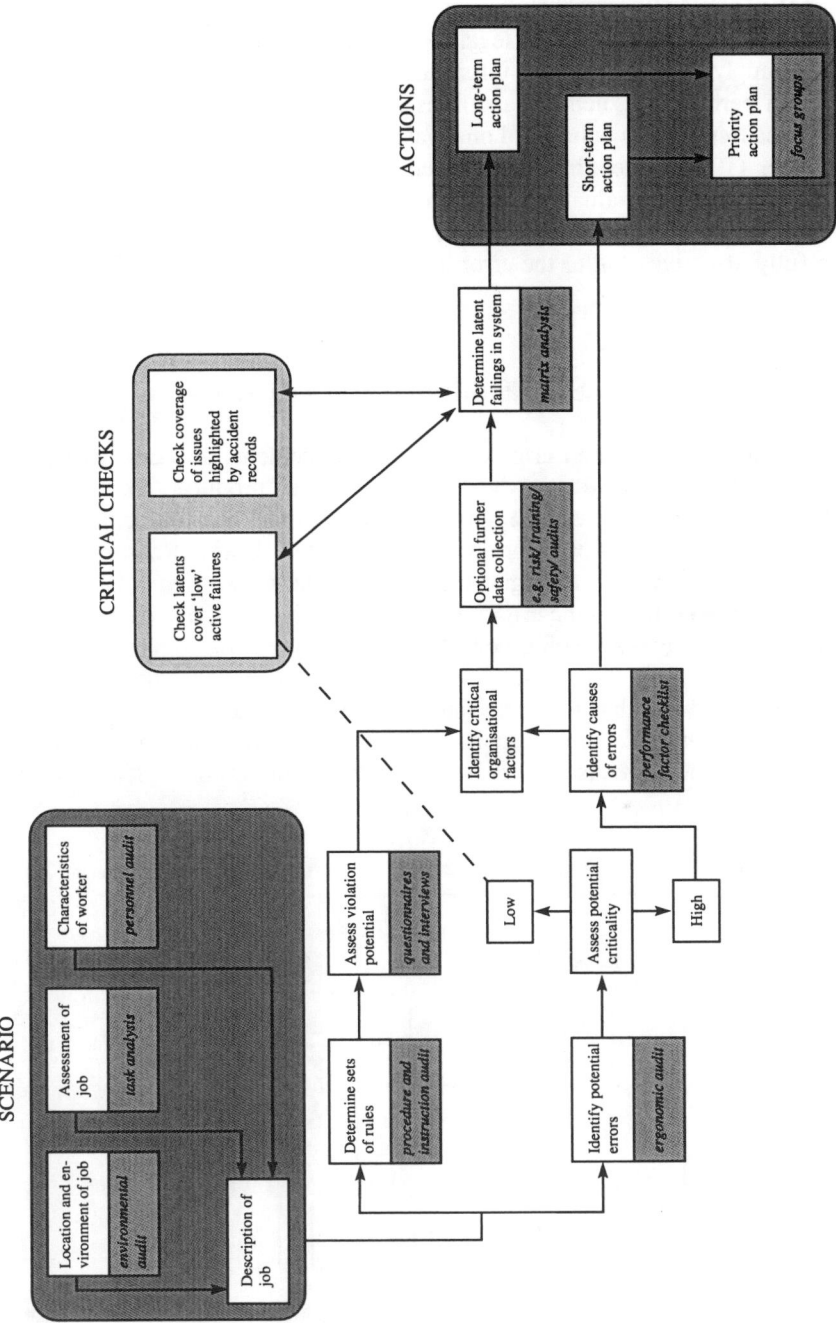

**Figure 3.2** The primary analysis lines in the BeSafe approach (Benedyk and Minister, 1995).

One interesting feature is that accident statistics from the organisation are used only as an ancillary check, not as the fundamental data indicating human error. Simpson's view is that the legal form of reporting of accidents does not in general supply sufficient helpful information for this depth of analysis, with the exception perhaps of scientific enquiries into  large-scale disasters, and even then the causes may be concluded on legal rather than scientific grounds.

Bamber (1986) states that, for accident prevention, one must identify risks so that they can be evaluated and controlled *before* accidents happen. Safety evaluation therefore needs to be carried out in a predictive way if prevention is to be fully effective – after the error has occurred it is too late. BeSafe allows this predictive evaluation.

## 3.4  Applying BeSafe to Product Safety Evaluation

In the same way as human error contributes to accidents at work, so human error contributes to accidents when using consumer products. In both cases there is an interaction between a person and a technical environment which is subject to design decisions and behavioural rules. Using Reason's model (Figures 3.1 and 3.3), we might suggest that the *person/team* is now the *product user* (who makes the error), the *workplace* is now the *use environment* (which sets up error-producing conditions), and the *organisation* is now the *design environment and societal framework* (which take decisions and influence procedures that feed into latent and active failure pathways). There are similar *defences* set up, such as design standards, procedures, or safety interlocks; but there are also weaknesses in such defences, not the least because

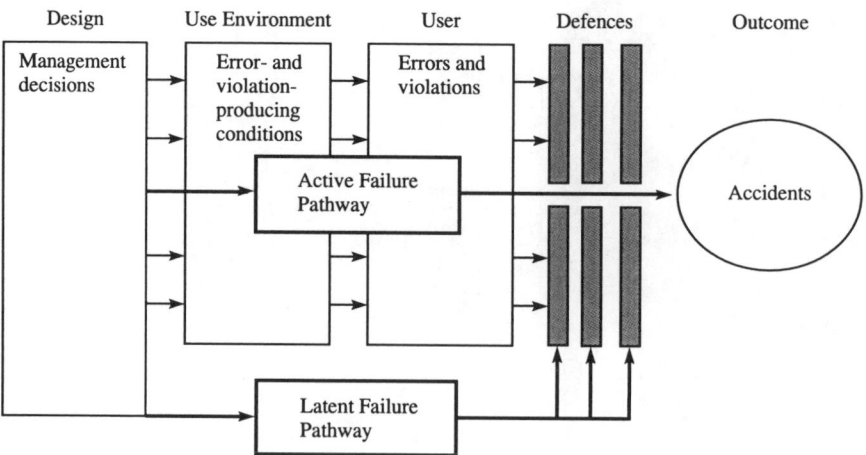

**Figure 3.3**  Reason's model applied to products.

there is no adequate means to foresee which people will use products, how they will use them and where they will use them (Kanis and Weegels, 1990). Thus, accidents do happen.

Take a familiar product example, the electric lawnmower. There are many accidents to consumers using lawnmowers (Consumer Safety Unit, 1983). Through the application of hazard analysis (using methods such as that of Benel and Pain, 1985) product designers have attempted to prevent the most obvious risks, using for example safety switches, and standards have been brought in for aspects such as the blade rotation stopping time (British Standards Institution, 1974). These are reasonable defences in the system. However, users have been known to tape up the safety switches to override them, and to disobey instructions not to disentangle the blades with the circuit connected: both such behaviours have led to accidents (Fulton, 1982). If we are to ensure that a lawnmower is actually as safe as possible, we must evaluate not only the ergonomics of the human/machine system, but the basis of the original design decisions, the reality of the use environment, and the nature of the user's risk perception. We need to identify the capacity for both errors and violations, and ultimately block both the active and latent failure pathways.

It is proposed that the BeSafe Method could be adapted for product safety evaluation, by redefining the *Scenario* from Figure 3.2 as in Figure 3.4. Instead of a job to be evaluated, it is the use of a product. A full description of this use requires information about the target population of users, recognising that there will be a range represented rather than just the single worker on the job in the industrial case. A number of population characteristics will be needed, such as age, anthropometry, experience and so on. Information is also required about

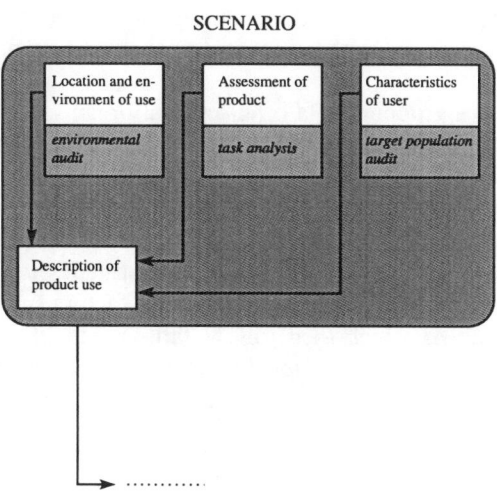

**Figure 3.4**   The Scenario for product safety evaluation using BeSafe.

the physical and societal environment of use, especially as, typically in product use, this may be variable and uncontrolled by any standard. Lastly, there must be full information about the product design itself and its intended function.

To illustrate, taking our lawnmower example, a sample of such information might be: the fact (*location*) that the use environment (the garden) is remote from the source of electricity; the fact (*environment*) that the work area is uneven and not necessarily free of obstacles; the fact (*task*) that, by nature of its function, a lawnmower has exposed cutting blades; the fact (*user*) that not just adults use the product (even though designers may have intended the product for adult use).

Placing this *Scenario* at the beginning of the BeSafe flowchart in Figure 3.2, this is followed by the two primary analysis lines in the BeSafe method: the ergonomic audit, and the instructional and behavioural audit. The ergonomic audit of the product and the performance factors which follow it are searching for all the design features that predispose human error when the product is used in its intended fashion. General examples might be: limitations of reaches or strength requirements, ambiguity or absence of information display, confusing controls, or a difficult environment such as darkness. The list of checks is long and thorough. It is important to identify all the potential errors in the system of product use. In the case of the lawnmower, we might learn that, for example: the power switch setting is ambiguous, so that the user cannot easily tell if it is on or off; or, the reach to the safety switch is too great for a ten year old's hand; or, the machine is unstable so that the blades are often exposed when turning on rough ground.

The second analysis line in BeSafe is the instructional and behavioural audit, which attempts to identify all the factors (individual or social) that predispose violations. Product users must be questioned here to get at their motivations for their behaviours. In the case of the lawnmower, users might report that they do not shut off the circuit when disentangling the blades because they resent walking all the way back to the electrical socket to disconnect, for example (Fulton, 1982).

BeSafe requires a grouping of all the 'critical organisational factors' at this point. In product evaluation, we take a looser view of this 'organisation'. The standards committees and the product designer, manufacturer, distributor, maintainer and advertiser, throughout the product lifecycle, are all part of the organisation, as are the product user's peer group, family or colleagues if they influence the user's capabilities, behaviour and attitudes towards the product. The final capacity for human error is fed by many factors from many sources.

The BeSafe analysis which then follows allows identification of the latent failings in the system, and the development of action plans to redress the weaknesses. (The reader is referred to Simpson *et al.*, 1997, for details of the method.) Such plans for actions to improve the lawnmower example might include designing an integral circuit-breaker into the machine, ensuring that the design is tested on young people and minimising maintenance requirements through good design features.

```
 1  Industrial/commercial

    1.1  For employees
         1.1.1  Installed      e.g. lathe; spot-welding tool
         1.1.2  Portable       e.g. shovel; fork-lift truck
    1.2  For customers/visitors
         1.2.1  Installed      e.g. cashpoint; vending machine
         1.2.2  Portable       e.g. shopping trolley; hire tools


 2  Domestic/leisure

    2.1  Public
         2.1.1  Installed      e.g. playground slide; drinking fountain
         2.1.2  Portable       e.g. communal dustbin; lakeside lifebelt
    2.2  Individual
         2.2.1  Installed      e.g. dishwasher; home computer
         2.2.2  Portable       e.g. lawnmower; pushchair
```

**Figure 3.5**   Taxonomy of products.

The scope of the action plans tends to vary according, for example, to the environment of work and the types of worker involved, and so it is with products. Figure 3.5 introduces a product taxonomy developed for the purpose which has proved useful in distinguishing products according to these characteristics, especially in discussions of action plans. It is helpful to differentiate the products first along the industrial/domestic divide, then according to group or individual use. Lastly, there are some differences between products that are stationary because they are installed, and those whose location of use varies because they are movable or portable. All the sections apart from 1.1 would normally be regarded as consumer products, although some are provided for the public by commercial concerns, and others are privately owned by the user. Their 'consumer' tag originates from the type of user, i.e. the public, rather than from the type of owner or the product's purpose.

## 3.5   Limitations of BeSafe

There are two prime areas in the BeSafe method for products where the model from industry is weakest; these are the *Scenario* data collection at the beginning, and the *action plans* at the end.

The data collection in the consumer product environment involves a much more wide-ranging and uncertain scenario than BeSafe deals with in industry. In industry, one can measure a particular worker in an observable environment, following known procedures. In product use, the users are variable, the use environment is variable (and may even be unknown) and many factors are diffuse or impractical to measure. The product use itself may be irregular (is a table for sitting at to eat or for standing on when changing a light bulb?). For these reasons it may be necessary to express data in terms of probabilities rather than certainties.

Management strategies for action are developed at the end of the BeSafe evaluation to address (primarily latent) failings in the system. However, the management of the product usage environment tends to be difficult. Many controlling measures used to prevent accidents in industry do not apply in this environment. For example, it is rare to be able to introduce training for users of products; supervision is usually absent; selection of users is impossible; warnings to users cannot be backed up with disciplinary action; instructions are often not read or unheeded; maintenance is often scanty and uncontrollable. This is well illustrated in Table 3.1 where there is a clear trend from the fully industrial and installed environment on the left to the fully domestic and portable environment on the right, of decreasing opportunities for ergonomic and other intervention to enhance safe usage of the product.

This leaves a small list of possible strategies that can be enacted to enhance the consumer product's 'safety culture'. At the top of this list are design features of the product itself – and of the product environment (e.g. installation) if that is accessible and appropriate – which afford safe behaviour by the user. This approach is positive both from the standpoint of ergonomics and from that of product safety standards which state that all hazards should, in principle, be eliminated by design (Van Weperen, 1993). For example, it is critical to relate  the product features to the physical and mental requirements of the particular users. Thus, in the case of our lawnmower, it is essential to understand that children do use this product, and that if the controls do not fit young hands, this will be inviting unsafe behaviour by the users. Lawnmower controls should be designed to fit and be worked safely by the full range of the *actual* or *probable* user population.

Another part of the design approach will be to strengthen the 'defences' in the system through good design. For example, make it as impossible as can be to bypass a safety mechanism on a product such as a lawnmower. Include 'forcing functions' (Norman, 1988) which have been tested in the field on a full range of users and proved effective. One useful route is to allocate some safety features away from the user and into the product mechanism, so that the load on the user is reduced and the temptation to violate is lessened. A relevant example here is that of the hairdryer, which is dangerous to use near water such as in bathrooms. Yet the bathroom is a tempting place for the drying of wet hair. For hotel bathrooms, therefore, the designer separated out the dangerous electrical parts and fixed them to the wall so that the risk was minimised, but the hotel guest can still use the hose of the dryer in the wet environment. Many products also demonstrate that features allocated to the mechanism only benefit the user, in terms of ease of learning and ease of use, as well as removing potential human error. An example of this is the computer diskette, designed so that it can only be inserted into the drive in one way: the correct way.

**Table 3.1** Feasibility of intervention for product categories

| | Product categories (see Taxonomy, Figure 3.5) | | | | | | | |
|---|---|---|---|---|---|---|---|---|
| | Industrial/ employees/ installed (*e.g. lathe*) | Industrial/ employees/ portable (*e.g. shovel*) | Industrial/ visitors/ installed (*e.g. cashpoint*) | Industrial/ visitors/ portable (*e.g. shopping trolley*) | Domestic/ public/ installed (*e.g. slide*) | Domestic/ public/ portable (*e.g. dustbin*) | Domestic/ individual/ installed (*e.g. dishwasher*) | Domestic/ individual/ portable (*e.g. lawnmower*) |
| Training | F | F | I | I | I | I | I | I |
| Selection | F | F | I | I | I | I | I | I |
| Supervision | F | M | M | I | M | I | I | I |
| Sanctions | F | F | M | M | I | I | I | I |
| Instructions | F | F | M | M | I | I | F | F |
| Labelling | F | F | F | F | M | M | M | M |
| Design of product | F | F | F | F | F | F | F | F |
| Design of environment | F | M | F | I | M | I | I | I |
| Maintenance | F | F | F | M | M | M | M | I |

F = Feasible;  M = Moderately difficult;  I = Impossible

Also open to enhancement are instructions, labels, warnings and help systems on products. It is essential to maximise the efficacy of this product information (Van Weperen, 1993), while recognising that some users will still ignore it completely or find it unusable. The objective is to minimise, through appropriate design, the proportion of users tempted or forced to overlook it. The presented information must be timely, visible, useful, usable and intelligible (by the widest possible range of potential users). It needs to be tested under field conditions, under conditions of wear and low maintenance, and under realistic task scenarios.

A third avenue open for support, in the case of consumer products, is the product image. There is room for much more safety awareness in the marketing and advertising of products, and this is supported by the design profession, driven by liability legislation concerns (Dodge, 1987). Advertising may be the only form of training that the consumer is exposed to, and if it misrepresents the importance of safety in product use, either by omission or commission, it opens the way for potential human errors. To illustrate, when child cyclists are shown on television adverts (whether for bicycles or not), if the children are seen to be wearing cycle safety helmets this enhances the viewers' awareness and acceptance of the safe way to use the bicycle.

Lastly, where training is available, it must instil a safety culture into the user, using validated educational methods. This includes self-training packages, most commonly found in consumer products such as home computers. In addition, a demonstration of the consequences of a human error, under safe training conditions, is often helpful. Any car driver who has been given experience of a skid-pan learns better how to avoid handling errors that could lead to an accident. Of course, there are still weaknesses in this approach, not the least that consumer products are often used by those other than the original trained user. Training information, unlike design features such as affordances, cannot be handed on to the next user.

In conclusion, it is the view of the authors that because there are limitations in management strategies in the area of consumer products, those strategies remaining assume vastly increased importance in the prevention of human error. In recognition of the underlying latent failures in the use system, it becomes necessary to 'over-design' the ergonomics of the product.

An illustration of the efficacy of this new approach to product safety evaluation is given in the following case study.

## 3.6   Case Study: Playground Slides

A study was set up to evaluate the utility of the BeSafe method in a consumer product environment (Minister and Benedyk, 1995). The environment chosen was the public children's playground and its associated equipment such as slides. Many accidents occur in playgrounds when using such equipment –

54 000 playground accidents were treated in hospitals in Britain in 1989 (Department of Trade and Industry, 1990). Despite the introduction of standards such as BS 5696 (British Standards Institution, 1979), accidents in playgrounds still account for approximately 6 per cent of child injuries (Safety in Playgrounds Action Group, 1994).

A playground should provide a child with an enjoyable play experience; one which is inviting, stimulating and exciting. Critically, it should provide a child with a safe place to play. Risk reduction in this environment is therefore important. The slide is, however, a consumer product, accessible to all, and methods for risk reduction are not straightforward, as illustrated in Table 3.1.

An evaluation of safety of slides in public playgrounds was carried out and the BeSafe method for assessing and reducing risk was applied to this environment (Simpson, 1994). Safety assessments using ergonomic techniques were carried out in the following areas: the playground equipment, the environment of the equipment, the attitudes of the users and their risk perceptions, and the interaction of the users with the equipment. In particular, children and their supervising adults were questioned about likely violations of use (e.g. walking up a slide instead of sliding down it), and individual slides were rated for 'safety value' to indicate how risky each might be under expected usage, and for 'play value' (Heseltine, 1988) to estimate how likely children might be to misuse the equipment through boredom. Further details of the method are reported in previous papers (Benedyk and Minister, 1995; Minister and Benedyk, 1995). Relevant parts of the BeSafe method were applied to the results. A selection of the slides data is presented in Figure 3.6, linked to the BeSafe flowchart to illustrate its application.

The data indicated that one of the behaviours most likely to cause an accident was that of climbing up the slope of a slide. It was found that children have a lower perception of risk involved in such behaviours than the adults who supervise them. However, where children had their own experiences of accidents to themselves or their peers, a raising of risk awareness was observed. Moreover, it was evident that children frequently violate despite an awareness of the risks involved, probably owing to bravado or group pressure.

The following were found to be the common latent failings in playgrounds (see Figures 3.7–3.10), and action was considered necessary in all of the areas to improve their safety:

■ Failure to mend/replace damaged and vandalised equipment
■ Failure to mend/replace/repaint old, worn or rusty equipment
■ Faults in layout (especially placing equipment for younger and older children close together)
■ Low play value provided
■ Installation faults
■ Entrapment points in equipment

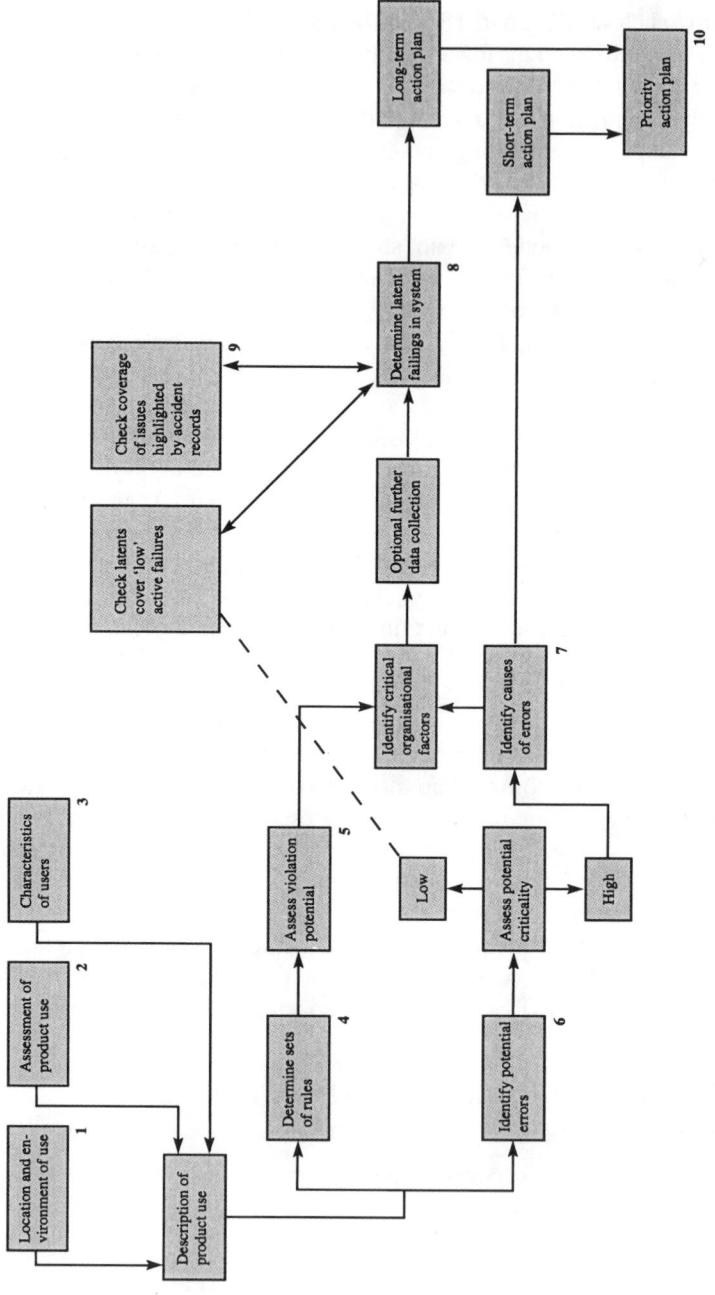

**Figure 3.6** Slides, the most common playground equipment item, are used to illustrate the BeSafe approach for consumer products (Benedyk and Minister, 1995).

# Key:

**1 Location and environment**  Slide of particular design; in playground of measured play value; sited on particular surface; positioned relative to other equipment

**2 Product use**  Queue for turn, climb steps holding onto rail, wait till chute is clear, sit and push off, bring feet down at bottom, stand and move quickly away to allow next user on slide

**3 User characteristics**  Children of any size, ability and experience from toddler to full-grown teenager need to be catered for; generally unsupervised; often in crowds

**4 Sets of rules**  E.g. do not walk up the slope of a slide; do not push others at the top of the slide; slide down one person at a time

**5 Violation potential**  E.g. walking up slide slope observed in 50% of sample; 33% of users say they do it frequently; judged dangerous by parents and supervisors but not by users; both highly risky and highly likely to occur

**6 Potential errors**  E.g. shooting off bottom of slide; falling from top of slide; slipping on steps and falling onto sharp stump of broken step

**7 Causes of errors**  E.g. lack of run-off (due to poor design of slide); slides which are too high with no protective structure (due to poor design); broken and missing steps (due to poor maintenance)

**8 Latent failings**  E.g. poor design; poor maintenance; low supervision; low play value of this and other items in the same playground

**9 Accident issues**  High tower slides are responsible for severe and fatal injuries to children due to falling

**10 Action plan**  Improve design to enhance play value, fit range of users better, and protect against high falls; ensure regular checks on maintenance; remove unsafe equipment; assess value of supervision; attempt awareness campaign among children

- Very high equipment with no protection against falls
- Design of equipment that encourages misuse by children
- Weak instructional training and supervision by adults.

Playgrounds and their environments can be seen to have many of the problems of consumer products in general, where optimisation of the interaction between the user and the equipment is concerned. Typically, playground users cover a wide range of abilities, ages, sizes and attitudes, management is remote and weak, and maintenance is generally haphazard. Many of the factors which could commonly be optimised to improve the interaction between users and equipment are inaccessible in this particular environment: this is shown clearly in Table 3.1.

It is thus true of slides that instructions and warnings are of limited value, since many of the users cannot read them and the rest generally choose to ignore them. The product must be open to all, so user selection is irrelevant, and it is not possible to control user behaviour with rewards for good and safe behaviour and sanctions for bad or dangerous behaviour. Even where design is concerned, playgrounds have an additional problem that workplaces do not

**Figure 3.7**  Violation – walking up a slide.

**Figure 3.8** Violation — several users on a slide at once.

**Figure 3.9** Poor maintenance — broken step with sharp point.

generally have to deal with, and that is the vast diversity of ability, skills and sizes of body parts of the users. Formal supervision is rare and costly, and training of users of the equipment (the children) is sparse, informal and impossible to measure.

It would seem, then, that approaches usually open to ergonomists for risk reduction measures are limited in this environment, and the only measures that can be relied upon are ergonomic optimisation of design (taking direct account of probable user behaviour) and, to a certain extent, maintenance of the

**Figure 3.10**    Poor design – steep slide with no run-off.

equipment and the environment. Ensuring the safety of children using slides in playgrounds must therefore primarily depend upon over-designing safety into the equipment and its layout and upon maintaining it properly. For example, slides built into the side of a slope are much safer in the case of falls than are tower slides, yet provide the same play value.

The BeSafe method is found to have a practical application in this context, although some of its elements are less appropriate than when used in organisations. However, it can be used to demonstrate where steps can be taken in general to improve safety in the use of slides. From Figure 3.6, it is clear that using BeSafe helps structure the safety evaluation of the product, helps to ensure that the approach is an ergonomic one and is comprehensive, and helps target the appropriate methods for reduction of risk in product use. These are benefits that could find application in a wide range of other consumer products. Thus, in principle, BeSafe acts not just as a safety evaluation method, but also shows the way forward to enhance product safety through an ergonomic approach to the reduction of potential human error.

## Acknowledgement

The support of International Mining Consultants Ltd, which owns the IPR of BeSafe, is gratefully acknowledged. Any interest in the use or application of BeSafe should be addressed to the Ergonomics and Safety Management Unit, International Mining Consultants Ltd, Bretby Business Park, Ashby Road, Burton-on-Trent DE15 0QD, UK.

# References

BAMBER, L. (1986) Principles of accident prevention. In Ridley, J. (ed.), *Safety at Work*. Sevenoaks: Butterworths.

BENEDYK, R. and MINISTER, S. (1995) Applying the BeSafe method to evaluation of product safety. In Stanton, N. (ed.), *Ergonomics in Consumer Product Design and Evaluation*. Loughborough: Ergonomics Society.

BENEL, D. C. R. and PAIN, R. F. (1985) The Human Factors Usability Laboratory in product evaluation. In *Proceedings of the Human Factors Society 29th Annual Meeting*. California: Human Factors Society.

BRITISH STANDARDS INSTITUTION (1974) *Specification For Powered Lawnmowers*, BS 5107. London: HMSO.

BRITISH STANDARDS INSTITUTION (1979) *Play Equipment Intended For Permanent Installation Outdoors*, BS 5696. London: HMSO.

CONSUMER SAFETY UNIT (1983) *Powered Tool Accidents*. London: Department of Trade and Industry/HMSO.

DEPARTMENT OF TRADE AND INDUSTRY (1990) *Playgrounds: A Summary of Accidents in Public Playgrounds*. London: HMSO.

DEPARTMENT OF TRADE AND INDUSTRY (1994) *General Product Safety Regulations*, Guidance document. London: HMSO.

DODGE, D. A. (1987) Product safety and liability prevention. In *Interface 87: Human Implications of Product Design, Proceedings of the 5th Symposium on Human Factors and Industrial Design in Consumer Products*. California: Human Factors Society.

ERGONOMIC AND SAFETY MANAGEMENT UNIT (1995) *The BeSafe Training Support Manual*. Burton-on-Trent: International Mining Consultants Ltd.

FULTON, E. J. (1982) *Lawnmower Accident Investigation*. Loughborough: Institute for Consumer Ergonomics.

HEALTH AND SAFETY EXECUTIVE (1992) *Provision and Use of Work Equipment Regulations*, Guidance document. London: HMSO.

HESELTINE, P. (1988) *Playing Safe – A Checklist for Assessing Children's Playgrounds*. London: National Playing Fields Association and Research Institute for Consumer Affairs.

KANIS, H. and WEEGELS, M. F. (1990) Research into accidents as a design tool, *Ergonomics*, **33**(4).

MCCLELLAND, I. (1995) Product assessment and user trials. In Wilson, J. R. and Corlett, E. N. (eds), *Evaluation of Human Work* (2nd edition). London: Taylor and Francis.

MINISTER, S. and BENEDYK, R. (1995) The utility of the potential human error audit in an ergonomic evaluation of public playground safety. In Robertson, S. A. (ed.), *Contemporary Ergonomics 1995*, London: Taylor & Francis.

NORMAN, D. A. (1988) *The Psychology of Everyday Things*. New York: Basic Books.

PHEASANT, S. (1991) *Ergonomics, Work and Health*. London: Macmillan.

REASON, J. (1995) A systems approach to organisational error, *Ergonomics*, **38**(8).

RYAN, R. P. (1987a) Consumer behaviour considerations in product design. In *Proceedings of the Human Factors Society 31st Annual Meeting*. California: Human Factors Society.

RYAN, R. P. (1987b) Hazard analysis for safe consumer product design. In *Interface 87: Human Implications of Product Design, Proceedings of the 5th Symposium on Human Factors and Industrial Design in Consumer Products*. California: Human Factors Society.

SAFETY IN PLAYGROUNDS ACTION GROUP (1994) personal communication.

SCHOONE-HARMSEN, M. (1990) A design method for product safety, *Ergonomics*, **33**(4).

SIMPSON, G. C. (1994) Promoting safety improvements via potential human error audits. In *Proceedings of the 25th Conference of Safety in Mines Research Institutes*. Pretoria, South Africa: SA Chamber of Mines.

SIMPSON, G. C., TALBOT, C. F. and RUSHWORTH, A. M. (1997) BeSafe: from knobs and dials to safety management. Personal communication.

STOOP, J. (1987) The role of safety in the design process. In Wilson, J. R., Corlett, E. N. and Manenica, I. (eds), *New Methods in Applied Ergonomics*. London: Taylor & Francis.

VAN WEPEREN, W. (1993) Guidelines for the development of safety-related standards for consumer products, *Accident Analysis and Prevention*, **25**(1).

# CHAPTER FOUR

# A systems analysis of consumer products

NEVILLE STANTON[1] and CHRISTOPHER BABER[2]

[1]*Department of Psychology, University of Southampton*
[2]*Industrial Ergonomics Group, University of Birmingham*

## 4.1 Predicting Human Error

We are all familiar with the annoyance of errors that we make with everyday devices, such as switching on an empty kettle, or making mistakes in the programming sequence with a video cassette recorder. People have a tendency to blame themselves for 'human error'. However, the use and abuse of the term has led to authors questioning the very notion of human error (Wagenaar and Groeneweg, 1988). 'Human error' is often invoked in the absence of technological explanations. One can argue that human error is not a simple matter of one individual making one mistake, so much as the product of a design which has permitted the existence and continuation of specific activities which could lead to errors (Reason, 1990). Predicting human error may strike the reader, at first sight, as implausible. However, if we know an activity that is to be performed, and the characteristics of the product being used, then it should be possible to indicate the types of error which may arise. This is the gist of all methods aimed at predicting human error: first define what actions need to be performed, and then indicate how these actions might fail. Techniques have been developed for the detailed and systematic assessment of a person's activities. A structured approach enables an analyst to identify potential points in tasks where errors could have significant negative consequences. From this assessment, preventive strategies can be sought to minimise the consequences or reduce the likelihood of error. An abundance of methods for identifying human error exist, some of which may be appropriate for the analysis of consumer products. In general, most of the existing techniques have two key problems. The first of these problems relates to the lack of representation of the external environment or objects. Typically, human error analysis techniques do not represent the activity of the device and material that the human interacts with, in more than a passing manner. Hollnagel (1993) emphasises that human reliability analysis (HRA) often fails to take adequate account of the context in which performance occurs. The second major problem of existing techniques is that there tends to be a good deal of dependence made upon the judgement of the analyst. Different analysts, with different experience, may make different predictions regarding the same problem (called inter-analyst reliability). Similarly, the same analyst may make different judgements on different occasions (intra-analyst reliability). This subjectivity of analysis weakens the confidence that can be placed in any predictions made.

Typically, human error identification (HEI) techniques are not related to psychological literature. This may be no bad thing (if the techniques work), but ultimately it may limit their ability to predict or otherwise deal with human performance. A distinct shortcoming is the lack of any cognitive analysis in HEI. Most techniques are unable to look beyond observable, physical operations (e.g. pressing buttons), to higher levels of performance (e.g. decision making). Although cognitive psychology has yet to deliver a fully validated, unifying theory, it does offer some insight into human thought processes (Newell, 1990). This literature has been largely ignored in the development of HEI techniques (which are often modifications of existing engineering approaches).

There is also a lack of verification of the methods. Indeed, before the study by Kirwan (1992), no systematic comparison of HEI techniques had been reported in the literature. In general, ergonomics methods tend to be poorly validated, and HEI is no exception. Based upon these arguments, it is suggested that any new HEI technique needs to learn from the shortcomings of past techniques. At present, most developments seem only to involve minor changes of existing techniques. Certainly, ease of use is a commendable facet to any HEI technique, but techniques should also be based upon an account of human performance within the context of the task and the product used, if they are to deal with *human* error.

## 4.2 A Systems Approach

In predicting human error it would seem sensible to consider how humans and devices interact. The foundation for such an approach relies upon general systems theory (von Bertalanffy, 1950). This theory is potentially useful in addressing the interaction between subcomponents in systems (i.e. the human and the device). It also assumes a hierarchical order of system components, i.e. all structures and functions are ordered by their relation to other structures and functions, and any particular object or event comprises lesser objects and events. General systems theory describes a system in dynamic terms:

- Activity results from continual adaptation of the system components.
- Changes in one component affect other components and the whole system.
- Systems components become linked by exchanges (inputs and outputs).
- Within a component an internal conversion occurs.
- Each type of input has a corresponding output.
- Errors become apparent at boundaries between components.

The input–conversion–output cycle of a human/machine system is illustrated in Figure 4.1. Information regarding the status of the machine is received by the human part of the system through sensory and perceptual processes and converted to physical activity in the form of input to the machine. The input modifies the internal state of the machine and feedback is provided to the human in the form of output.

Of particular interest here is the boundary between humans and machines, as this is where errors become apparent. Figure 4.1 indicates both the exchange and conversion processes assumed within general systems theory. We believe that it is essential for a method of error prediction to examine explicitly the nature of the interaction. Many methods appear to do this in an implicit way, but leave consideration of the interaction to the judgement of the analyst. Whilst expert analysts, with a good understanding of the domain, may be able to do this task satisfactorily on most occasions, it does not make the analysis as objective as it could be. Thus, there is a need for a structured HEI technique which can account for this interaction. Interaction between humans and machines has been described previously as a form of problem-solving exercise (Baber and Stanton, 1992).

In order to study the cognitive and planning processes involved in problem solving, researchers tend to follow one of three traditions. First, planning is

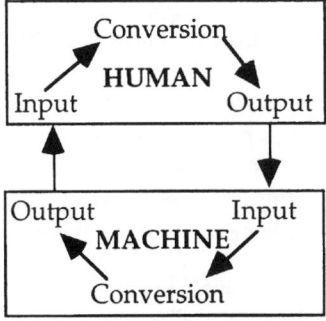

**Figure 4.1**   Input–conversion–output cycle of human/machine interaction.

seen as a hierarchical process which controls and guides the order and sequences of task performance. This is not dissimilar to a computer program. The implementation of the plan requires continual monitoring of the environment and adaptation of the plans (Miller *et al.*, 1960). Second, planning is seen as arising from a formal analysis of the problem domain, with plans being sets of procedures to move from one state to the other (Newell and Simon, 1972). Recent research has tended to emphasis of feedback in plan execution (Hayes-Roth and Hayes-Roth, 1979; Hoc, 1988). The third tradition extends the emphasis placed upon context, to suggest that planning is largely dependent upon environmental cues (Suchman, 1987).

The approach adopted in this chapter can be viewed as an amalgam of the three traditions. General goals and expectations enable plans to be developed in advance of interaction. These plans may be modified as interaction progresses, as performance is influenced by feedback. However, there may be situations in which an entirely new course of action is taken, where the original plans are hijacked and new courses of action take over. Reason (1990) calls this process 'thematic vagabonding'. We offer this account here as a unified problem-solving approach, where the planning process is modified by the context in which it operates.

Problem solving may be considered analogous to moving through a maze, from the initial state towards the goal state (Newell and Simon, 1972). Each junction has various paths, of which one is selected. Moving along a new path changes the present state. Selection of a path is equivalent to the application of a number of possible state-transforming operations (called operators). Operators define the 'legal' moves in a problem-solving exercise, and restrict 'illegal' moves or actions under specific conditions. Therefore, a problem may be defined by many states and operators, and problem solving consists of moving efficiently from our initial state to our goal state by selecting the appropriate operators (as illustrated in the Figure 4.2). Moving from states 1 to 4 presents the problem solver with future possible states of 11, 12 and 13. Had a different operator been applied, different states would result (e.g. 2 or 3).

When people move between states they also change their knowledge of the problem. Newell and Simon (1972) proposed that problem-solving behaviour can be viewed as the production of knowledge states by the application of mental operators, moving from an initial state to a goal state. They suggested that problem-solvers hold knowledge states in working memory, and operators in long-term memory. The problem solver then attempts to reduce the difference between the initial state and the goal state by selecting intermediary states (subgoals) and selecting appropriate operators to achieve these. Newell and Simon suggest that people move between the subgoal states by:

- noting the difference between present state and goal state;
- creating a subgoal to reduce the difference;
- selecting an operator to achieve this subgoal.

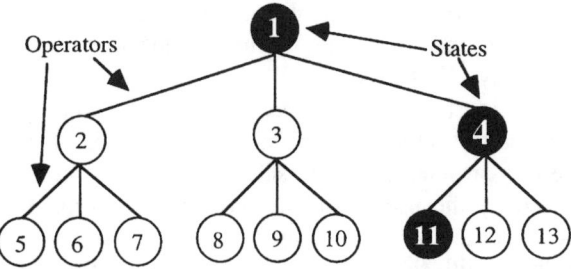

**Figure 4.2**   States and operators.

In this instance, conversion is influenced by the selection of an operator to achieve a subgoal. Thus it would appear that the cognitive demand of the task is substantially reduced by breaking the problem down, moving towards the goal in a series of small steps. In some cases, the machine output might be sufficient to suggest an appropriate operator. Gaver (1991) argues that information regarding the potential of a device is supplied not only by perception but also by exploration. Information about the operation of a device becomes apparent through interaction with that device. As the interaction progresses, new (previously unapparent) operators are made available. For example, a door handle affords the action of grabbing; once grabbed, the handle affords the action of turning (which releases the catch), this in turn affords the action of pulling the door open. Each action in turn leads to a new possibility that ultimately satisfies the goal of opening the door. Suchman (1987) refers to this type of activity as 'situated' planning. Errors can occur when there is an incompatibility between operators suggested by a device and its actual properties in two main ways. The first type of incompatibility is when a device does not actually allow an action that it appears to afford (e.g. a door handle that cannot be turned). The second type of incompatibility is when a device allows an action but results in an unexpected outcome (e.g. a door which is pushed rather than pulled). This idea accords with the problem-solving model of Newell and Simon (see Figure 4.2 for an example of sequential problems solving: moving from states 1 to 4 to 11) and general systems theory, for considering how one might examine the interaction between humans and machines, with the ultimate aim of predicting error.

## 4.3   Rewritable Routines

In order to move from current to relevant states, eliminating other possible states, the user needs to retain some record of the interaction and to have some means of assigning relevance to states. At each state, this record will be modified. Thus, it will need to be rewritable. We assume that the record will be held in working memory, presumably in the articulatory loop which has a limited duration (around 2 seconds). This means that unless the record is updated, it will decay. As the record will also guide the next action, we see this as a rewritable routine (Baber and Stanton, 1997). To some extent this notion is similar to the 'partial provisional planning' hypothesis of Young and Simon (1987). Figure 4.3 presents a simple schematic of this process. The possible states (interpreted by the user from the product) are compared against states which could lead to the goal. The comparator has a two-way connection to the rewritable routines (with the routines both influencing the comparator, i.e. by defining relevance, and taking the output to define action).

It is proposed that the routines are held in working memory. This raises the question of how the routines are formed and what other memory is used. Bainbridge (1992) presents a description based on interdependencies in working storage, which is modified in Figure 4.4. Typically, interaction with technology will require this structure to be completely rebuilt, particularly when confronted with an entirely new design of product. Further, even performance on well known products may impact on only part of the description, e.g. repeated performance may lead to a reduction in the range of actions available, rather than changes in the prediction processes. If this is the case, the planning ahead or acting beyond the current state will be difficult.

If interaction with technology has little need for prediction, then the process will involve step by step processing of states and planning at each state, i.e. situated activity (Suchman, 1987) or opportunistic planning (Norman, 1988). The argument is that interaction with public technology primarily involves defining available actions and comparing these with user goals. The notion of available action relates to the idea of an interface as a resource for action (Payne, 1991), in which the options presented to the user suggest what the user can do. This has been referred to as 'affordance' by Norman (1988). As previously discussed, Gaver (1991) suggests that affordances can be sequential.

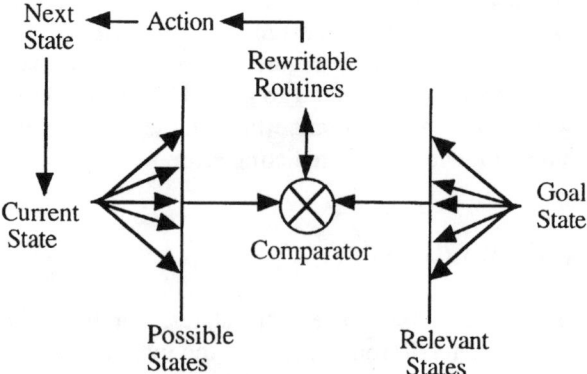

**Figure 4.3**　Simple schematic of rewritable routines (from Baber and Stanton, 1997).

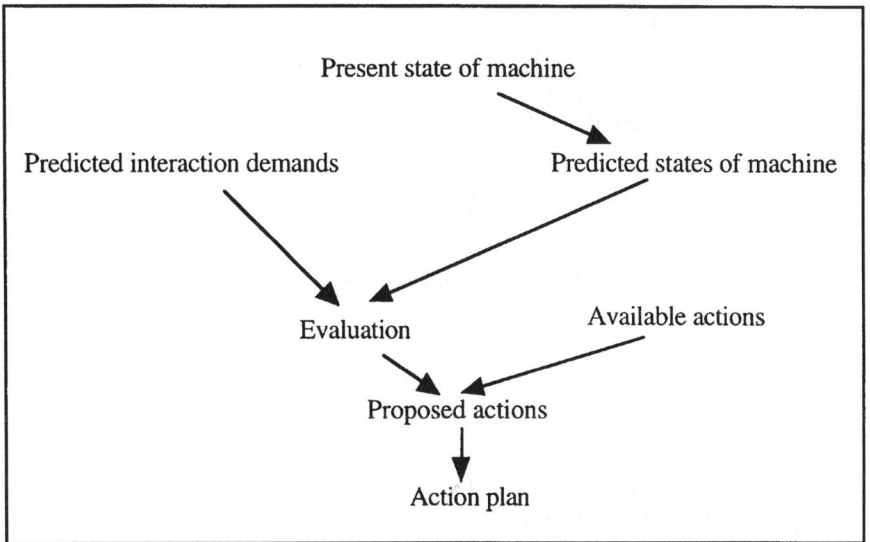

**Figure 4.4**   Interdependencies between working storage (from Bainbridge, 1992, p. 129).

The notion of available action will also relate to the idea of the user generating some local action specification, e.g. some belief that once action $i$ has been performed it will be necessary to perform action $j$ (if actions other than $j$ are required, this will be confusing). This could be like the notion of adjacency pairs in linguistics (an adjacency pair is defined by commonly occurring pairs of linguistic acts, such as question–answer). Are there adjacency pairs in human/machine interaction, i.e. local pairs of actions which always belong together. Returning to the door handle example, not only is there a sequence of affordances but there is also a sequence of tasks: grip handle + turn, turn handle + (door/lock action) + move door. There appear to be several points in an interaction at which users 'know' an action has failed: users know an action has failed when nothing happens, i.e. when outcome feedback does not appear; users know an action has failed when there is no action feedback (touching a touchscreen and there is no beep); users know an action has failed when they cannot perform the 'adjacent' action. The adjacent action will be determined by knowledge of similar machines, feedback presented by the product and user's interpretation of the feedback, i.e. its relevance. We have been working on an approach that attempts to capture the essence of this theoretical description of human activity and translates it into a methodology that can be used for predicting human error in product use. This methodology is called task analysis for error identification (TAFEI).

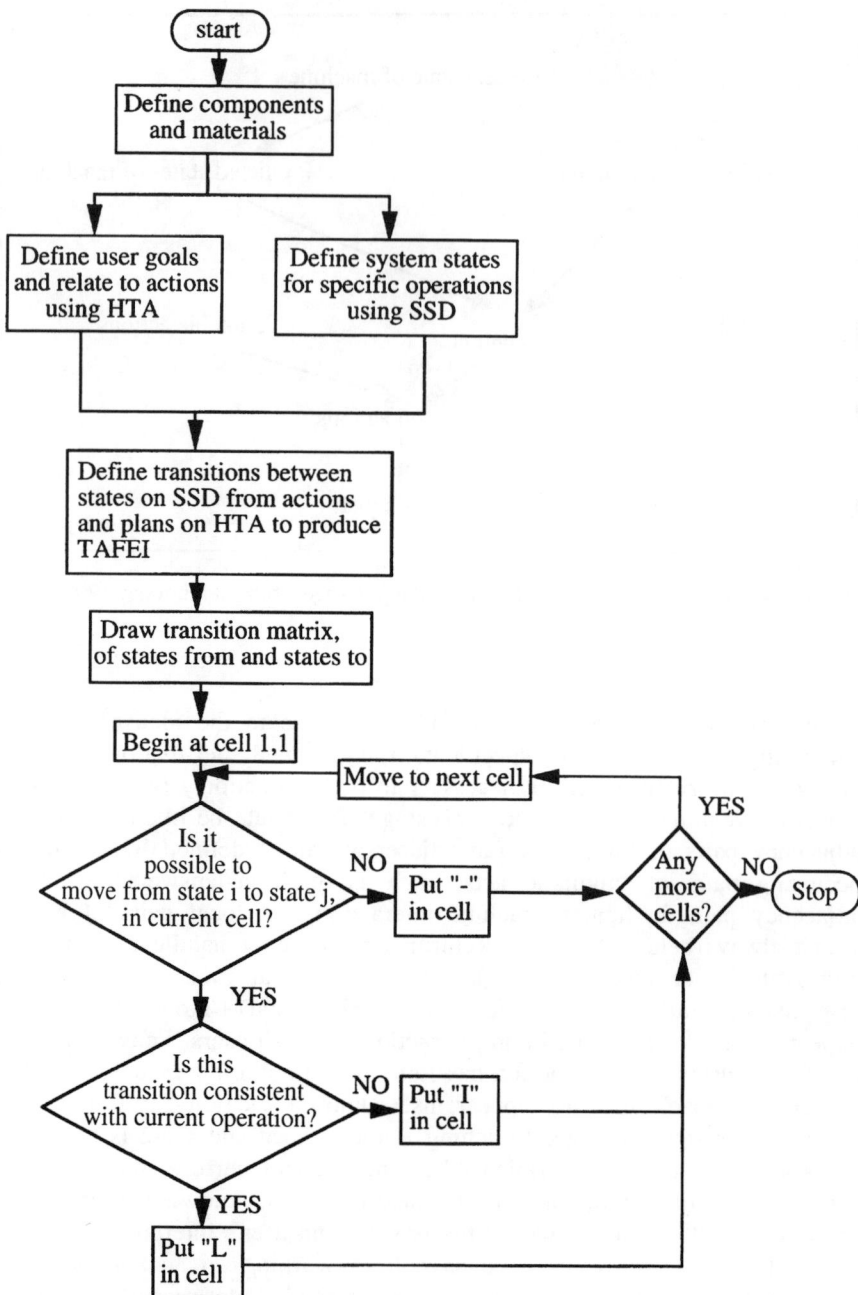

**Figure 4.5** TAFEI procedure.

## 4.4  Task Analysis for Error Identification

TAFEI (Stanton and Baber, 1991, 1993, 1996; Baber and Stanton, 1991, 1992, 1994) explicitly analyses the *interaction* between people and machines. TAFEI analysis is concerned with task-based scenarios and entails mapping human activity onto machine states. TAFEI analysis consists of three principal components: hierarchical task analysis (HTA), state/space diagrams (SSDs) and transition matrices (TMs). HTA provides a description of human activity, SSDs provide a description of machine activity and TMs provide a mechanism for determining potential erroneous activity through the interaction of the human and the device. In a similar manner to Newell and Simon (1972), legal and illegal operators (called *transitions* in the TAFEI methodology) are identified. This scenario analysis enables the analyst to consider different forms of human/product interaction. The TAFEI procedure is summarised in the flow diagram illustrated in Figure 4.5.

First, the system to be addressed needs to be defined. Next, the human activities and machine states are described in separate analyses. The basic building blocks are hierarchical task analysis (for describing human activity) and state/space diagrams (for describing product activity). These two types of analysis are then combined to produce the TAFEI description of human/ machine interaction. It is worth pointing out that the state/space diagram also has the potential to contain information about hazards or byproducts associated with particular states. From the TAFEI diagram, a transition matrix is compiled and each transition is scrutinised. Each transition is classified as impossible, illegal or legal, until all transitions have been analysed. Finally, illegal transitions are addressed in turn as potential errors, to consider changes that may be introduced. TAFEI has been used in the evaluation of a wide variety of applications, such as wordprocessors, automatic teller machines, in-car devices, ticket machines and vending machines. The two case studies presented here are based upon analysis of consumer products, both previously reported by Baber and Stanton (1994).

As mentioned earlier, people make errors even with relatively simple products such as domestic kettles. The first case study is based on an ordinary electric kettle. The HTA shown in Figure 4.6 describes the actions and goals associated with successfully boiling the kettle, but does not begin to indicate where problems might arise. Figure 4.7 shows the SSD, indicating those states that are necessary and sufficient for the kettle to be boiled. Only by mapping the activities from Figure 4.6 onto Figure 4.7 can the analyst begin to analyse the interaction between the two subsystems comprising human and product. This is shown in Figure 4.8 – the TAFEI diagram. In the TAFEI diagram the transitions between product states have been labelled from the HTA to show which human activities lead to changes in product states to reflect error-free performance.

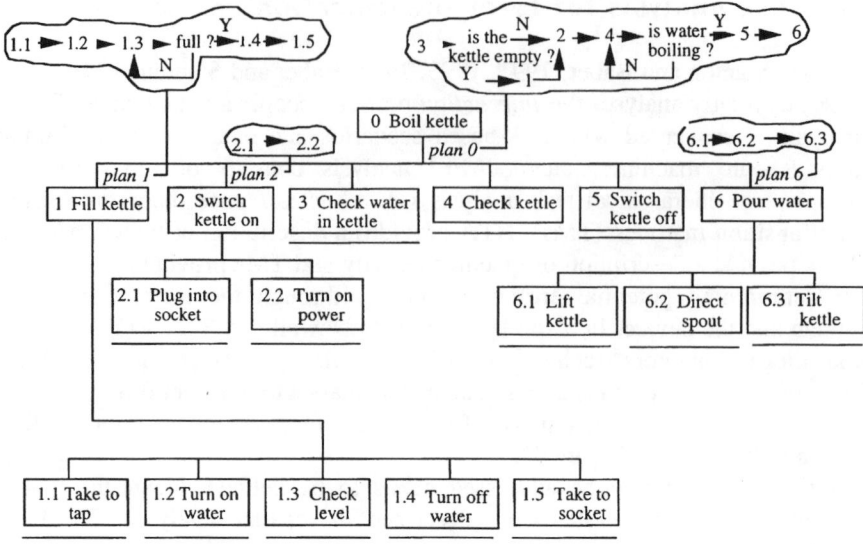

**Figure 4.6** HTA of kettle operation.

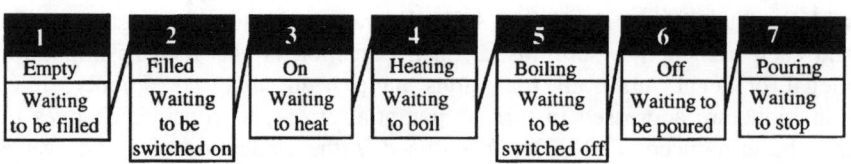

**Figure 4.7** SSD of kettle operation.

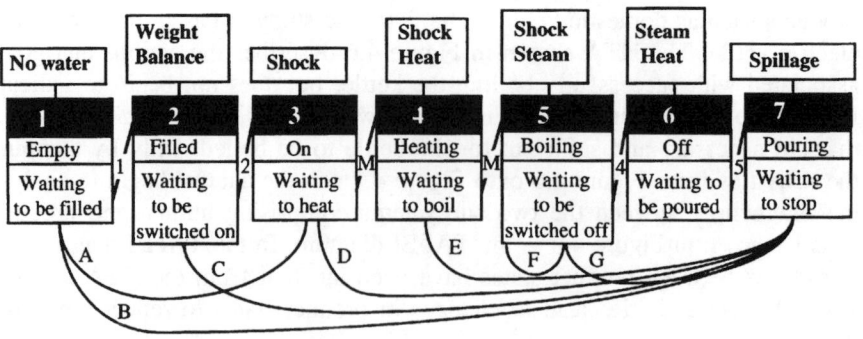

**Figure 4.8** TAFEI diagram of kettle operation.

From the TAFEI diagram we can begin to analyse where human errors might occur in the operation of the kettle. We have assumed Murphy's law: that is, if an error can occur then it probably will. Figure 4.8 not only shows the error-free path to successful performance, but also indicates where erroneous transitions may occur. The erroneous transitions have been labelled A to G. As a kettle is a relatively simple device with few states, it is easy to identify the erroneous transitions on the TAFEI diagram. However, in more complex devices it may be necessary to use a transition matrix, as shown in Figure 4.9.

In the transition matrix each cell has been categorised as legal (L), illegal (I) or impossible (−). It is worth pointing out that transitions 3 to 4 and 4 to 5 are performed by the kettle and therefore have been labelled M (for machine). The illegal transitions (or errors) have been identified as:

A. Switching on an empty kettle.

B. Attempting to pour water from an empty kettle.

C. Pouring out cold water.

D. Pouring water before it is hot.

E. Pouring water before it has boiled.

F. Not switching off a boiling kettle (noted as a recursive transition).

G. Pouring water before the kettle has been switched off.

We believe that the list of illegal transitions provides the designer with a focus for redesign of the product, to turn them into impossible transition. For example, a thermostatic switch could be used to switch the kettle off automatically, thereby eliminating the possibility of error F. This might also reduce the likelihood of errors D and E. A floating ball linked to a switch could prevent

To
state:

|        |   | 1 | 2 | 3 | 4 | 5 | 6 | 7 |
|--------|---|---|---|---|---|---|---|---|
|        | 1 | — | L | I / A | — | — | — | I / B |
|        | 2 | — | — | L | — | — | — | I / C |
| From   | 3 | — | — | — | M | — | — | I / D |
| state: | 4 | — | — | — | — | M | — | I / E |
|        | 5 | — | — | — | — | I / F | L | I / G |
|        | 6 | — | — | — | — | — | — | L |
|        | 7 | — | — | — | — | — | — | L |

**Figure 4.9**  Transition matrix for kettle operation.

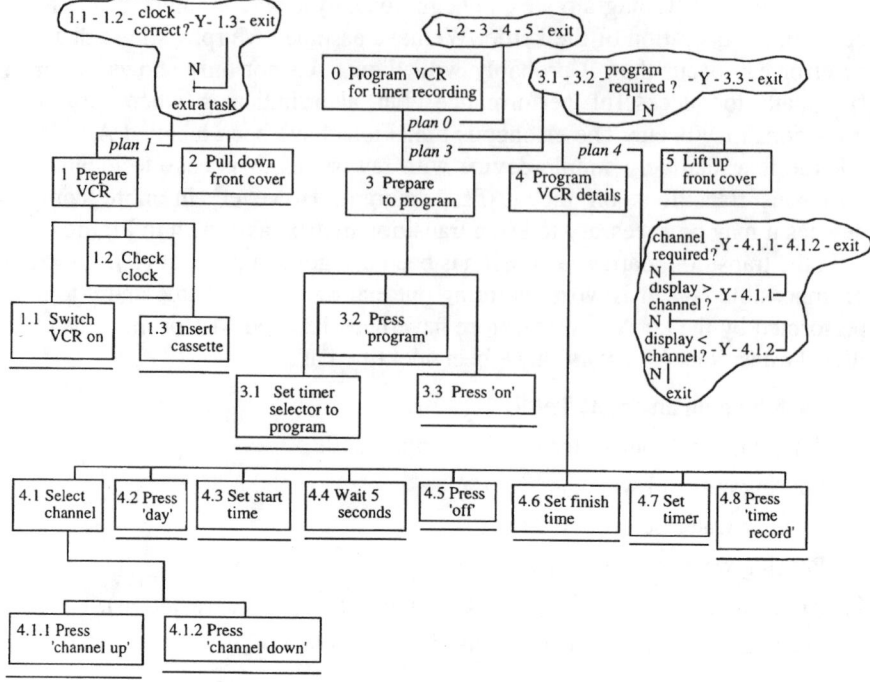

**Figure 4.10**   HTA of VCR operation.

the kettle being switched on whilst it is empty (error A). A clear plastic kettle might also help reduce the likelihood of errors A and B. In this way, design can be focused on providing solutions to potential user problems.

In the second example, the case of programming a video cassette recorder (VCR) is examined. This has become a notoriously difficult task and we think that a TAFEI analysis can help us understand why this is so. This analysis includes only part of the operation of a VCR: that of setting the VCR to record a single programme at a time in the future. The HTA, SSD and TAFEI diagrams are shown in Figures 4.10, 4.11 and 4.12, respectively.

The requirements for the transition matrix become apparent for managing larger state transition analyses. The transition matrix for VCR operation is shown in Figure 4.13.

As in the previous example, each cell has be categorised legal, illegal and impossible, as appropriate. Thirteen of the transitions are defined as illegal; these can be reduced to a subset of six basic error types:

A.  Switch VCR off inadvertently.

B.  Insert cassette into machine when switched off.

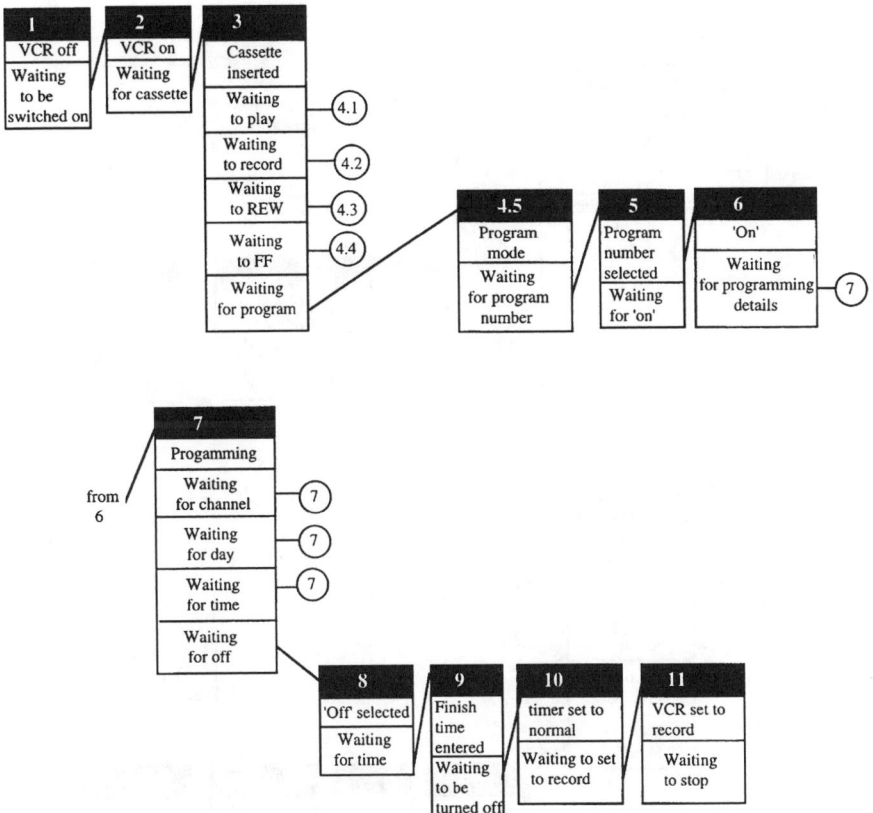

**Figure 4.11**   SSD of VCR operation.

C.  Program without cassette inserted.

D.  Fail to select programme number.

E.  Fail to wait for 'on' light.

F.  Fail to enter programming information.

In addition, one legal transition has been highlighted because it requires a recursive activity to be performed. These activities seem to be particularly prone to errors of omission. These predictions then serve as a basis for the designer to address the redesign of the VCR. A number of illegal transitions could be dealt with fairly easily by considering the use of modes in the operation of the device, such as switching off the VCR without stopping the tape and pressing play without inserting the tape. As with the previous example, the point of the analysis is to help guide design effort to make the product error-tolerant.

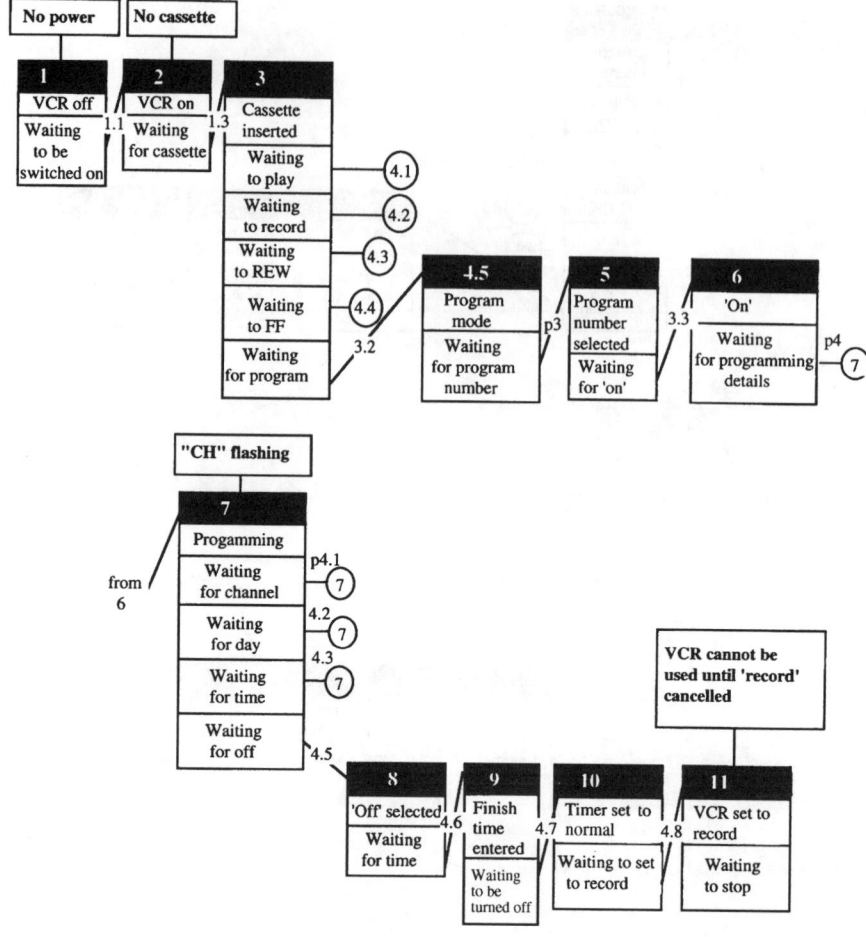

**Figure 4.12** TAFEI diagram of VCR operation.

## 4.5 Conclusions

The TAFEI method has developed into a useful approach for analysing human/ product activity. In addition to the growing number of case studies showing the application of TAFEI, we have also been conducting research studies into the reliability and validity of the approach. The results of these studies are encouraging, leading us to believe that the method is not only a useful adjunct to design activity but also a reasonable predictor of real performance. We are also hearing of reports of the TAFEI approach being used with success in a variety of product-based applications. This offers additional support as to its usefulness.

To
state:

|  | 1 | 2 | 3 | 4.5 | 5 | 6 | 7 | 8 | 9 | 10 | 11 |
|---|---|---|---|---|---|---|---|---|---|---|---|
| 1 | - | L | I | - | - | - | - | - | - | - | - |
| 2 | L | - | L | I | - | - | - | - | - | - | - |
| 3 | L | - | - | L | - | - | - | - | - | - | - |
| 4.5 | I | - | - | - | L | I | - | - | - | - | - |
| 5 | I | - | - | - | - | L | I | I | - | - | - |
| 6 | I | - | - | - | - | - | L | I | - | - | - |
| 7 | I | - | - | - | - | - | L | L | - | - | - |
| 8 | I | - | - | - | - | - | - | - | L | - | - |
| 9 | I | - | - | - | - | - | - | - | - | L | - |
| 10 | I | - | - | - | - | - | - | - | - | - | L |
| 11 | - | - | - | - | - | - | - | - | - | - | - |

From state:

**Figure 4.13** Transition matrix for VCR operation.

# References

ANNETT, J., DUNCAN, K. D., STAMMERS, R. B. and GRAY, M. J. (1971) *Task Analysis*. Department of Employment Training Information Paper 6. London: HMSO.

BABER, C. and STANTON, N. A. (1991) Task analysis for error identification: towards a methodology for identifying human error. In: Lovesey, E. J. (ed.), *Contemporary Ergonomics*. London: Taylor & Francis.

BABER, C. and STANTON, N. A. (1992) Defining problems in VCR use. In: Lovesey, E. J. (ed.), *Contemporary Ergonomics*. London: Taylor & Francis.

BABER, C. and STANTON, N. A. (1994) Task analysis for error identification: a methodology for designing error tolerant consumer products, *Ergonomics*, **37** (11) 1923–41.

BABER, C. and STANTON, N. A. (1997) Rewritable routines in human interaction with public technology. *International Journal of Cognitive Ergonomics*. In press.

BAINBRIDGE, L. (1992) Mental models and cognitive skill: the example of industrial process operation. In: Rogers, Y., Rutherford, A. and Bibby, P. A. (eds), *Models in the Mind*. London: Academic Press.

GAVER, W. W. (1991) Technological Affordances. In: *Proceedings of CHI'91, New Orleans, Louisiana, April 28–May 2, 1991*. New York: ACM.

HAYES-ROTH, B. and HAYES-ROTH, F. (1979) A cognitive model of planning, *Cognitive Science*, **3**, 275–310.

HOC, J-M. (1988) *The Cognitive Psychology of Planning*. London: Academic Press.

HOLLNAGEL, E. (1993) *Human Reliability Analysis: Context and Control*. London: Academic Press.

KIRWAN, B. (1992) Human error identification in human reliability assessment. Part 2: detailed comparison of techniques. *Applied Ergonomics*, **23**, 371–81.

MILLER, G. H., GALANTER, R. and PRIBRAM, K. (1960) *Plans and the Structure of Behaviour*. New York: Holt, Rinehart and Winston.

NEWELL, A. (1990) *Unified Theories of Cognition*. Cambridge, Mass.: Harvard University Press.

NEWELL, A. and SIMON, H. A. (1972) *Human Problem Solving*. Englewood Cliffs, NJ: Prentice Hall.

NORMAN, D. A. (1988) *The Psychology of Everyday Things*. New York: Basic Books.

PAYNE, S. J. (1991) A descriptive study of mental models, *Behaviour and Information Technology*, **10**, 3–21.

REASON, J. (1990) *Human Error*. Cambridge: Cambridge University Press.

STANTON, N. A. and BABER, C. (1991) A comparison of two word processors using Task Analysis For Error Identification. Paper presented at the HCI'91 annual conference at Herriot-Watt University, Edinburgh.

STANTON, N. A. and BABER, C. (1993) Task analysis for error identification. Paper presented to the joint MRC/RNPRC/APRC Workshop on Task Analysis, 5–6 April, Scarman House, University of Warwick.

STANTON, N. A. and BABER, C. (1996) A systems approach to human error identification. *Safety Science*, **22** (1–3), 215–28.

SUCHMAN, L. (1987) *Plans and Situated Actions: The Problems of Human/Machine Communication*. Cambridge: Cambridge University Press.

VON BERTALANFFY, L. (1950) The theory of open systems in physics and biology, *Science*, **111**, 23–9.

WAGENAAR, W. A. and GROENEWEG, J. (1988) Accidents at sea: multiple causes, impossible consequences. *International Journal of Man–Machine Studies*, **27**, 587–98.

YOUNG, R. M. and SIMON, T. (1987) 'Planning in the context of human–computer interaction'. In: Diaper, D. and Winder, R. (eds), *People and Computers III*. Cambridge: Cambridge University Press.

# CHAPTER FIVE

# Ergonomics and the evaluation of consumer products: surveys of evaluation practices

CHRISTOPHER BABER and MUNA G. MIRZA

*Industrial Ergonomics Group, University of Birmingham*

## 5.1 Introduction

In this chapter, attention is given to 'white goods', i.e., domestic products which are used in the preparation, storage, cooking, etc. of food. In order to consider how ergonomics can contribute to the evaluation of consumer products, two surveys were undertaken. In one survey, laboratories involved in the technical evaluation of white goods were approached. Personnel completed questionnaires concerning their work and the evaluation that they performed. In the second survey, ergonomics laboratories were approached and personnel involved in product evaluation provided information on their approaches to evaluation.

One might feel that it is difficult to separate evaluation from product design, and that good design practice should incorporate early evaluation (Gould *et al.*, 1987). However, this notion can be said to represent a specific perspective on the nature of evaluation. Other perspectives are equally valid, e.g. considering evaluation as a means of testing whether a product meets a particular standard, or considering evaluation as a means of assessing whether a product caused a particular accident, or considering evaluation in terms of comparing the design, performance, etc. of different products. In these instance, evaluation is in essence divorced from the design process (although, of course, the results of the evaluation can be fed back into future design activity). As part of this study, a further aim was to identify the major reasons for undertaking evaluation of consumer products, in order to determine the most appropriate means for introducing ergonomics into the evaluation process.

The aims of the study reported in this chapter were threefold:

- To determine how the evaluation of white goods is conducted.
- To determine the role of ergonomics in the evaluation of white goods.
- To identify the types of ergonomics evaluation practice which are useful and which can be applied.

In order to develop an argument for the role of ergonomics in product evaluation, the paper begins with a discussion of accidents which occur as a result of the use of white goods.

### 5.1.1  White Goods and Accidents

Each year, around 250 000 accidents requiring hospital treatment occur in domestic kitchens (HASS, 1992). Taking out accidents caused by slipping or involving knives, many of the remaining accidents involve white goods. The figures are sufficiently high to imply that consumer products can represent a significant, if small, risk to users. For this reason alone, it would appear important to evaluate products in terms of their safety. However, the importance of evaluation is underlined by the numerous standards, both national and international, which require manufacturers to indicate that their products are fit both for human use and for the uses for which they have been designed.

Norman (1988) has argued that human interaction with any form of product is heavily influenced by the users' goals and by the users' interpretations of the product. This means that some of the problems that arise from the use of consumer products may well be the result of purposeful action on the part of the user, rather than simply the result of product malfunction (Baber and Stanton, 1994). For instance, the relationship between the size of a person's hand and the width of an aperture in a food blender only becomes an issue if users are going to place their hand in the aperture. If the aperture leads to a spinning blade, then one might believe that putting a hand in the aperture would be foolish to say the least. However, one can foresee situations in which such activity can arise from a deliberate intention, e.g. the user may wish to dislodge food stuck around the blade or to add other ingredients. Thus, unless prevented from doing so by, for instance, interlocks fitted on a cover over the aperture, users may attempt to use their hands to dislodge food or add ingredients. In these cases, the risk of injury increases as a result of purposeful activity. One might feel that these risks can be reduced by simply having standards which require manufacturers to fit interlocks on all consumer products to prevent such actions. However, not only would this prove an expensive measure, it may also not be necessary on all products. Alternatively, one might wish to standardise the depth and size of apertures. However, this might lead to people using other objects, such as a wooden spoon rather than their fingers, and thus rather than eliminating the risk, one would simply allow the activity to migrate along another behavioural pathway leading to a different set of risks, e.g. the risk of injury from flying splinters if the wooden spoon comes into contact with the rotating blade.

The point of this discussion is that the use of white goods can represent a risk for users, i.e. can lead to accidents, and that many of the risks arise from the uses to which people put the products. The implication is that the evaluation of white goods should consider how people will use products of specific designs. Without going into great detail of the wide variety of evaluation procedures, the following section considers the relationship between evaluation practice and what might be termed 'models of user behaviour'.

### 5.1.2 Evaluating White Goods

In order to satisfy standards, it is necessary to demonstrate that products will have minimal risk attached to their use. One can consider risks in terms of the electrical, mechanical, physical (etc.) characteristics of the product. Products can be tested in order to ascertain how they will perform over periods of time, how they will malfunction and whether the malfunctions can lead to potential risks to users. Hence, product evaluation is directed at proving that products are fit for human use. However, in many of the standard evaluation procedures, there is no requirement for 'typical users' to be involved in the evaluation. This raises an interesting question for both ergonomics and evaluation practice: if products are tested to be fit for human use, to what extent should one seek to define human use? In other words, what model of user behaviour is being employed in the evaluation procedures?

The ergonomics literature generally accepts the fact that the 'model' held by designers of a specific product will differ from that held by the user. In other words, the knowledge of how the product functions, the expectations of what functions the product can perform, and the ability to use the product to perform these functions will be far more advanced for a person who has been involved with the product throughout its development, than for someone encountering the product for the first time. One significant implication of this difference in knowledge is that designers often have difficulty in considering the range of problems that people could conceivably face when using the product. Furthermore, it is also generally accepted that designers possess models of users, e.g. assumptions of how users will behave and act when using particular products. Although there has been little, if any, attention given to the models of users held by evaluators, one wonders whether a similar phenomenon is at work in evaluation procedures; that is to say, people who undertake the evaluation of white goods, while not explicitly considering 'ergonomics', may still hold certain beliefs and assumptions concerning what would be 'normal' human activity when using a particular product. The question then becomes not simply should one consider users in product evaluation, but in what detail should human action be considered.

The primary focus of technical evaluation is on the integrity of the product. However, user safety is often an important consideration in such tests, e.g. when measuring the heat produced by an oven, one can measure the temperature reached inside the oven and the amount of heat lost from the outside of the oven. The amount of heat lost could be used as an index of energy efficiency, but also implies concern over risk of burns to users. Physical tests could require that there is no sharp edge on the product in order to reduce the risk of injury to people, or that the components of toys will not cause injury or lead to suffocation. In order to determine the levels of such risk, one needs to consider how people are going to be using the product. Thus, 'human use' may be considered implicit in a number of technical tests. This raises the question of whether the technical tests should be modified to make the 'model' of human use explicit. There are a number of approaches to this question. One could aim to introduce ergonomics into standards; one could seek to ensure that technical tests are performed by people with knowledge of ergonomics; one could pass products on to ergonomists for evaluation; or one could incorporate ergonomics methods into the testing procedure.

## 5.2  Survey One: Technical Testing and Evaluation of White Goods

Ten laboratories were initially selected for this exercise. They were taken from NAMAS Directory D3, and were selected for their involvement with white goods. For example, the laboratories surveyed evaluated a wide variety of white goods (see Table 5.1). Following an initial telephone survey, six laboratories were selected. The remaining four were unable to participate in the survey owing to lack of time, lack of available personnel or concerns over confidentiality. This means that the laboratories used were selected on the basis of their willingness to participate in the study rather than because they represented typical examples of the testing laboratories in the UK. However, from informal discussion with respondents during telephone debriefing and during subsequent semi-structured interviews, it is felt that the responses elicited from this survey are supported by other organisations.

In this study, 60 questionnaires were posted to six testing laboratories. A total of 39 questionnaires were returned, giving a response rate of 65 per cent (the individual response rate of laboratories ranged from 50 per cent to 100 per cent, with the mode being 60 per cent). The respondents were well qualified, from diploma to masters level, and had around six years' experience of product evaluation. The respondents were employed by their laboratories primarily to undertake technical testing and evaluation.

### 5.2.1  Findings

The main findings of the survey are presented in Table 5.1. Although more information was collected, this section contains data pertaining to technical and ergonomic evaluation. Of principal interest to this chapter are the reasons for conducting technical evaluations in the laboratories and reasons for not conducting ergonomic evaluation. Furthermore, the research is part of a wider project investigating evaluation methods, and consideration of these also features in this section.

*Reasons for Undertaking Evaluation*

Technical testing is undertaken at the laboratories surveyed for a variety of reasons, the main one being that manufacturers or distributors may be required to prove compliance with standards. In general, the standards used tend to be national or European, as opposed to international. The manufacturer or distributor will contact a laboratory and ask for the evaluation to be performed. The standards are stipulated by the client or defined by legal bodies, although occasionally the laboratory staff select additional standards to apply. In addition to evaluation against standards, the laboratories often become involved in accident investigation. Products which have been involved in accidents are presented to the laboratories in order to ascertain whether the accidents were caused by product design faults. A final set of evaluation activities involves comparative evaluation of products, either of a range of new designs or of products from different manufacturers.

It is interesting to note that laboratories may be involved in one or more of the evaluation practices. This suggests that the nature of evaluation will vary across the practices. For instance, while many tests for compliance with standards are laid down in standards documents, accident investigation and product comparison can require the development of novel evaluation methods and procedures. Furthermore, the client for evaluation practices will also vary. This will have implications for the nature of reporting, e.g. reports to the public versus reporting to an inquest. This suggests that evaluation practices will vary not only in terms of how they are approached but also in terms of how they are recorded and reported.

When products were tested against standards, 25 of the 39 respondents felt that products are 'reasonably safe' when they conform to standards. However, the remaining respondents felt that the application of standards does not, of itself, make products safer. A possible reason for this relates to the feeling that, even after the application of a standard, it is not possible to legislate for the

Table 5.1  Summary of findings for survey of technical evaluation practice

| | | | | Laboratory | | |
|---|---|---|---|---|---|---|
| | 1 | 2 | 3 | 4 | 5 | 6 |
| Products tested | Coffeemakers, Freezers, Washing machines, Elec. cookers | Kettles, Knives, Frying pans, Irons, Freezers, Coffeemakers, Washing machines, Blenders, Hotpots | Kettles, Knives, Frying pans, Heaters | Kettles, Knives, Frying pans, Irons, Elec. heaters | Kettles, Knives, Frying pans, Irons, Elec. heaters, Coffeemakers, Washing machines, Freezers, Hotpots | Kettles, Knives, Frying pans, Irons, Elec. heaters, Blenders |
| Qualifications of personnel | Diploma/degree/masters | Diploma/degree/masters | Diploma/degree/masters | Diploma/degree/masters | Diploma/degree/masters | Diploma/degree/masters |
| Technical tests undertaken | Mechanical Physical Inflammability Chemical Electrical | Mechanical Physical Inflammability Chemical Electrical | Mechanical Physical Inflammability Chemical Electrical Hygiene Radioactivity | Mechanical Physical Inflammability Chemical Electrical | Mechanical Physical Inflammability Chemical Electrical | Mechanical Physical Inflammability Chemical Electrical |
| Ergonomic tests undertaken | None | None | None | None | None | None |
| Standards employed | British EU (National standards of client/market) | British EU (National standards of client/market) | British EU (National standards of client/market) | British EU (National standards of client/market) | British EU (National standards of client/market) | British EU (National standards of client/market) |
| Reasons for evaluation | Comparative analysis | Legislation, Accident investigation, Product modification | Legislation | Legislation, Accident investigation | Legislation | Legislation, Accident investigation |
| Source of samples | High street | Client | Client | Client | Client | Client |
| Duration of tests | 12 min.–8 weeks | 10 min.–4 weeks | 1 min.–8 weeks | 5 min.–4 weeks | 5 min.–4 weeks | 5 min.–8 weeks |

behaviour of users and that people could still use a product in an unsafe manner.

## Selection of Samples

The laboratories receive samples of new products from the manufacturers and evaluate these in terms of prescribed testing procedures in order to determine whether the product meets the required standard. The samples could be new products or modifications of existing products which are being developed as new or replacement lines. In general, samples for evaluation are provided by clients. Occasionally, the client supplies a number of samples and personnel select samples from this population. In cases of accident investigation, the products may be supplied by legal bodies. It is interesting to note that product selection rarely involves purchasing products from the high street (although one laboratory obtains most of its samples in this manner).

## Nature of Evaluation

The laboratories tend to perform a similar range of tests, e.g. mechanical, physical, electrical, chemical and inflammability tests. The aims of these tests are not simply to test for product safety but also to examine durability and reliability, which may also impact on safety. One laboratory also included hygiene and radioactivity tests. The length of the evaluations varied enormously, from 1 minute to 8 weeks. The main factor determining the length of an evaluation was the complexity and range of the tests to be used.

## Role of Ergonomics

The respondents were asked whether they conducted any form of ergonomics evaluation in their laboratories. None did. The most common reasons for this were lack of knowledge and/or lack of training in ergonomics theory and methods (20 of the 39 respondents agreed with these statements). Eight of the respondents felt that the technical tests were sufficiently detailed to capture some of the ergonomic factors required in their evaluation. Four respondents felt that ergonomics testing was not required by their clients, and only two respondents believed that ergonomics testing could not be applied owing to pressure of time. A further point to note is that, according to the respondents, ergonomics evaluation is not a requirement in any of the standards used in the laboratories. When asked whether they felt that ergonomic evaluation would be beneficial, 23 of the respondents felt that it would, and 16 felt that it would not.

Respondents were also asked how they would prefer to perform ergonomic evaluation. The majority opted for expert appraisal, i.e. following the approach that they tend to use now. Only around one-quarter of the respondents felt that using people to perform user trials would be acceptable. The respondents recognised the need for persons undertaking ergonomics evaluation to be formally trained, either as ergonomists or as trained personnel in the laboratory. However, there was also a feeling that ergonomics information could be provided to aid evaluation, perhaps in the form of documentation.

### 5.2.2   Conclusions

There is no ergonomics evaluation in technical testing houses. There is little requirement for it by either the standards or the clients of the laboratories – manufacturers and retailers who are concerned to meet these standards. On the other hand, accident investigations are undertaken to help in product liability cases. Ergonomic considerations of product use do not seem to play a significant role in either situation. Products which are difficult to use, are unsafe or are not designed to minimise the risk of accident can lead to liability action, and one might feel that ergonomics evaluation could contribute to an understanding of these issues.

The personnel of technical testing do not feel that they have sufficient expertise to conduct the tests. The majority of the respondents opted for expert appraisal to perform ergonomic evaluation. This is not surprising in that it is what they are used to, but it is also of some concern in that so few of the respondents felt that running user trials would be useful. Further, in the absence of guidance or standards on the procedures for conducting these evaluations, one might anticipate differences in the manner in which data are collected and reported between different evaluators and between different laboratories. In order to determine what type of information would be required to enable personnel at technical testing houses to conduct ergonomics evaluation, a study of evaluation practice by ergonomists was conducted.

### 5.3   Survey Two: Ergonomics Evaluation Practice

One can propose four main methods which can be used in the ergonomics evaluation of consumer products:

- Descriptions of potential risk, e.g. using scenario analysis, fault tree analysis, hazard analysis.
- Descriptions of physical characteristics of users, e.g. using anthropometric/ biomechanic analysis.
- Descriptions of the product, e.g. using checklists or 'expert appraisal'.
- Descriptions of the use of the product, e.g. observing people using the product.

In broad terms, scenario analysis, fault tree analysis and hazard analysis are used to describe sequences of actions and the consequences of each action in the sequence, specifically in terms of risk to users. Anthropometric/ biomechanic analysis, on the other hand, is used primarily to examine the relationship between the design of the product and user characteristics, e.g. comparing the dimensions of adult and child hands with the size and depth of apertures, say, leading to sharp blades. Checklists and expert appraisal tend to focus on static aspects of the product and, while they can capture a great deal of information, they may not be as useful as observations of product use. Consequently, one might anticipate that ergonomics evaluation should include observations of people using products. Given the requirement for accountability in data collection under quality standards, such as ISO 9001 and BS 5750, it is also necessary to have some means of recording data in a consistent fashion.

### 5.3.1   Data Collection

Twenty-six people employed by six organisations were surveyed. The laboratories were selected from the Ergonomics Society Directory and were chosen, following a telephone survey, because they participated regularly in the evaluation of white goods. All respondents were practising ergonomists, with qualifications to degree or masters level in a human science, and had around 7 years' experience in product evaluation (ranging from 1 to 15 years).

### 5.3.2   Findings

Table 5.2 summarises some of the findings from this survey. Respondents were also asked about the methods that they used, and these results are discussed below.

*Reasons for Undertaking Evaluation*

All of the respondents felt that ergonomics was important in the evaluation of consumer products. Twenty-five of the respondents felt that the main benefit of ergonomics was that it allowed one to assess whether a product could be operated or maintained effectively by the user. Thus, the main aim of ergonomic assessment for the respondents would appear to be the measurement of the usability of products.

Twenty of the respondents felt that ergonomics evaluation was also undertaken to determine the design inadequacies in a product which would affect usability and safety, and 18 further saw this role as the resolution of usability problems. For one organisation at least, ergonomics was felt to offer commercial benefits in product design and marketing, and this influenced the number of comparative tests performed.

**Table 5.2** Summary of findings for survey of ergonomics evaluations practice

| | Laboratory | | | | | |
|---|---|---|---|---|---|---|
| | 7 | 8 | 9 | 10 | 11 | 12 |
| Qualifications of personnel | Diploma/degree/masters | Diploma/degree/masters | Diploma/degree/masters | Diploma/degree/masters | Diploma/degree/masters | Diploma/degree/masters |
| Technical tests undertaken | None | None | None | None | None | None |
| Reasons for evaluation | Usability, Product comparisons | Usability, Product comparisons | Usability, Product comparisons | Usability, Design faults, Product comparisons | Usability, Design faults, Product comparisons | Usability, Design faults, Product comparisons. Commercial advantage |
| User selection | Pool | Local | Local Agency Friends Employees | Pool | Pool Local | Pool Agency |
| Number of users | 1–5 | 1–5 | 11–15 | 16–20 | 6–10 | 11–15 |
| Duration of tests | 2–12 weeks | 12 hours–4 weeks | 1 hour–4 weeks | 4–12 weeks | 4–24 weeks | 4–12 weeks |
| Source of samples | Client | Client | Client | Client | Client High Street | Own products |

*Nature of Evaluation*

The ergonomists employed a variety of methods for evaluation. There were, however, two broad classes of method used: observations of people using the products, either in real time or from video recordings of individuals and pairs of users, and self-report, ranging from expert appraisal to checklists to questionnaires and interviews. Furthermore, there were many instances of two or more methods being used simultaneously. Respondents argued that combinations of methods allowed for data to be collected from different perspectives. The idea behind this seemed to be that data from different perspectives could converge to give a clear indication of problems. However, there was no mention of how the convergence would be performed or of what would happen if the data conflicted. Furthermore, it is interesting to note that some of the pairings may, in fact, be collecting the same sort of data, e.g. subjective opinion, simply in a different format, e.g. questionnaires and interviews. The most common pairings of methods were:

- Questionnaires and interviews
- Questionnaires and observation
- Observation and interviews
- Observations and checklists
- Questionnaires and rating/ranking
- Checklists and rating/ranking
- Expert appraisal and checklists
- Expert appraisal and videotaping of product use (either by individual users or by pairs of users).

The ergonomists were asked to rate the methods in terms of frequency of use, ease of use and utility. A simple regression was performed on these ratings, using a Spearmans rank order correlation. All of the correlations were above 0.8 (i.e. Utility × Ease of use = $\rho$ 0.878; Utility × Frequency of use = $\rho$ 0.884; Frequency of use × Ease of use = $\rho$ 0.93). The justification for the choice of method was not investigated in this study, although this is the focus of an ongoing research project.

In general, the ergonomics evaluation appeared to take in the order of 1 hour to 12 weeks, depending on the nature of the product and the evaluation method used. However, there were often severe time constraints on evaluation, imposed either by the client or by market pressures. The evaluations involved between 1 and 20 people, and the average appears to lie below 15. While this may not appear to be a particularly large sample size, research has suggested that one can discover a large proportion of potential problems/risks with around eight to ten users (Virzi, 1992; Nielsen and Landauer, 1993). Efforts were made to ensure that, where possible, users were drawn from a representative population, often through the use of employment agencies or from a pool held by the organisation.

Finally, the ergonomists were evenly split as to whether it would be sensible to include ergonomics in standards, with 13 saying yes and 13 saying no. The reasons for not including ergonomics in standards ranged from the difficulty of defining 'human behaviour' to the problems of allowing tests to be sufficiently flexible to deal with variations in products and users while being sufficiently well defined to allow for consistency between testers.

### 5.3.3  Conclusions

From the survey of ergonomists, one can see that there appears to be a limited number of methods which are used to perform evaluation and that, in general, these methods are combined in order to develop a more detailed picture of the use of a product. If one compares the methods used with the list of approaches above, then it is instructive to note that the laboratories surveyed tended towards descriptions of the product and of the use of the product. The main aim of evaluation appeared to be how easy a product was to use. The evaluations took from one hour to several months, and involved up to 20 users. All the ergonomists felt that observation of product use was essential.

## 5.4  Discussion

From these surveys, one can see a clear divide between technical and ergonomic testing of consumer products. The personnel involved in technical testing agreed that there might be some benefit in introducing ergonomics into evaluation, although not all were convinced. From the survey of ergonomists, it would appear that dynamic evaluation is important, i.e. studying the use of a product. It is instructive to note that some three-quarters of the respondents in the first study were not in favour of using such an approach.

It was interesting to note that the principal concern of the ergonomists in this survey was on the usability of the product, rather than on safety. One might suggest that, just as human use is implicit in many technical tests, so safety is implicit in many of the ergonomics tests. However, the ergonomists felt that ergonomics evaluation would lead to safer products, as well as products which were easier to use. If ergonomics is a matter of applying a broad body of knowledge through a collection of methods, then one might feel that training could permit personnel in technical houses to conduct ergonomics evaluation. However, the differences in background and experience between personnel in technical testing houses and ergonomists may be sufficient to lead to differences in evaluation practice. Further, the use of combined methods and the reliance on user testing by the ergonomists means that there will a number of methods which need to be considered. With the continuing development of ISO 9241 and related standards, one might anticipate usability becoming a significant issue in the evaluation of a range of products. This makes the question of ergonomics important in the evaluation of white goods.

## References

BABER, C. and STANTON, N. A. (1994) Task analysis for error identification: a methodology for designing error-tolerant consumer products, *Ergonomics*, **37**, 1923–42.

GOULD, J. D., BOIES, S. J., LEVY, S., RICHARDS, J. and SCHOONARD, J. (1987) The 1984 Olympic Message System: a test of behavioural principles of system design, *Communications of the ACM*, **30**, 758–69.

HASS (1992) *Home and Leisure Accident Research 1990 Data*, 14th Annual Report of the Home Accident Surveillance System. London: Dept of Trade and Industry, Consumer Safety Unit.

NIELSEN, J. and LANDAUER, T. K. (1993) A mathematical model of the finding of usability problems. In *Proceedings of the ACM Conference on Human Factors in Computing*. Seattle, WA: ACM, pp. 249–56.

NORMAN, D. A. (1988) *The Psychology of Everyday Things*. New York: Basic Books.

VIRZI, R. A. (1992) Refining the test phase of usability evaluation, *Human Factors*, **34**, 457–68.

# PART TWO

# Institutions Involved in Design and Evaluation of Consumer Products

# CHAPTER SIX

# Application of ergonomics and consumer feedback to product design at Whirlpool

ADRIAN MARTEL

*Whirlpool Europe s.r.l., Biandronno, Italy*

## 6.1 Introduction

Although it has only been known in Europe for a few years, Whirlpool is one of the world's leading domestic appliance manufacturers. Originally an American company, it entered the European market in 1992 by taking over the 'white goods' part of the Philips business. After an introductory period when products were marketed as Philips–Whirlpool, the company now markets under its own name, as two other brands – Bauknecht and Ignis – and as various 'private label' brands.

Various groups perform consumer research within Whirlpool, and this chapter focuses on methods used by the Centre for Applied Product Ergonomics (CAPE) which is a part of the Central Industrial Design (CID) department. The role of the ergonomist can take either a high or low profile in the product development depending on the project, as illustrated in Figure 6.1.

## 6.2 High-profile Ergonomic Development

The most innovative synergy between design and ergonomics is achieved through a close working relationship between the designer and ergonomist throughout the product development process. This can generally be divided into three phases:

■ Collecting information for the design

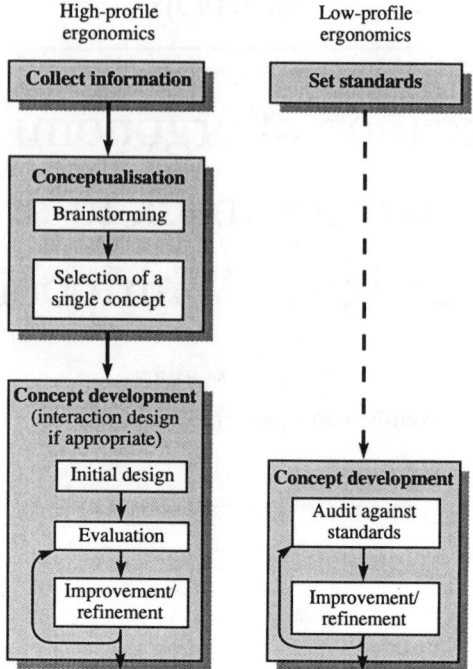

**Figure 6.1**   The high- or low-profile role of ergonomics in the design process.

- Conceptualisation
- Development of a single design concept.

### 6.2.1   Collecting Information for the Design

The first phase involves establishing the following:

- What consumers need the product to do.
- How previous versions of the product are perceived and how they can be improved.
- What competitors are doing which might be also beneficial to this design.
- Which ergonomic guidelines are relevant.

*Focus Groups*

A relatively fast and cheap way of discovering consumers needs for a product is to use focus groups – six to eight representative consumers at a time are invited to come to the usability lab to talk about use of the product of interest.

Participants are usually chosen who have experience of the type of discussion under development, though it can also be interesting to have one or two inexperienced users to add different perspectives.

The room in which the focus group is held is made as similar as possible to the real environment – in Whirlpool's case usually the kitchen. Sessions are conducted informally and typically begin with participants being asked, in turn, to talk about their current product and how they use it. They may also be asked what they wish they could do with their product and what aspects may be improved upon. This usually uncovers enough anecdotes to promote an interesting discussion covering the topics of interest to the mediator throughout the rest of the session. The mediator should:

- be prepared with a list of topics to be discussed and be flexible in the order in which they are covered, avoiding any disruption to the flow of the conversation. One way to open up a topic is by trying to expand upon a point which has been already made by somebody;
- make only brief notes because it is important that he or she should stay in touch and in control of what is being said and because writing can add disruptive pauses. Sessions are therefore recorded (audio with or without video) so that the detail can be noted later. Sometimes another member of staff is available to make notes 'live' during the session;
- prevent the conversation from getting sidetracked into issues less relevant to the research;
- make sure that everyone has a chance to express their own experiences and opinions: otherwise some participants can dominate discussions with the others simply agreeing with them;
- not express any opinion;
- try not to answer questions but rather to ask in return what the participant thinks the answer should be.

Later in the session it is sometimes useful to have a creative period where participants can draw and discuss their ideas. This also introduces a welcome element of variety and interest into the discussion.

*Home Observation*

Focus groups can only reveal consciously known and expressible needs. Another type of needs – 'latent' or unconscious needs – are better determined by interviewing and watching the behaviour of representative consumers using their products – if possible, in their own home. The needs are sometimes expressed as habits or 'coping strategies' developed by the user to overcome some difficulty. Home observation is more time-intensive than focus groups but can be more revealing because the user is in a real context.

*Perception of Previous Product*

Studies of the previous version of the product also provide useful information. One valuable source of information is the field test data conducted just before final release of the last version of the product onto the market. People use a preproduction version of the item at home for a period of weeks or months and record any difficulty they experience. This point in development is often too late for correcting many of the issues which may arise, and so it is only with the next version of the product that much of the information can be used.

Another useful source is information collected from known purchasers of the product, usually with postal questionnaires after around six months of ownership. The number of people contributing information may be quite high and, with the product being subject to prolonged real use, is highly valid.

Finally, this research may be supplemented by relatively quick testing at the usability laboratory if the specific information required is not available.

*Benchmarking*

It is important for a commercial company to know what its competitors are doing in order to help it remain competitive. Consumer preferences for aspects of competing designs are therefore studied using benchmarking.

Typically, a group of competitors' machines are assembled along with previous own brand machines, all with the brand markings hidden to minimise bias. Participants are asked to rate the features using both rating scales and open questions; this may be presented either as a verbal checklist or as a written questionnaire.

*Ergonomic Guidelines*

In addition to all the subjective research on consumer needs and preferences, it is useful also to collate relevant ergonomic guidelines generated both externally (published research in books and journals) and internally (previous tests). An organised system of information retrieval is needed, and Whirlpool uses a database called Recommendations for Ergonomic Design (RED). In updating this as an integral part of ergonomic research, information is always evolving and accessible. Brief ergonomic recommendations on any topic are presented with pictures of design examples (as shown in Figure 6.2), a detailed discussion and references to the source of the information – the latter two are mainly for the benefit of the ergonomist.

### 6.2.2  Conceptualisation

The next phase of the process applies this information and seeks to establish the design direction – the particular configuration and implementation of

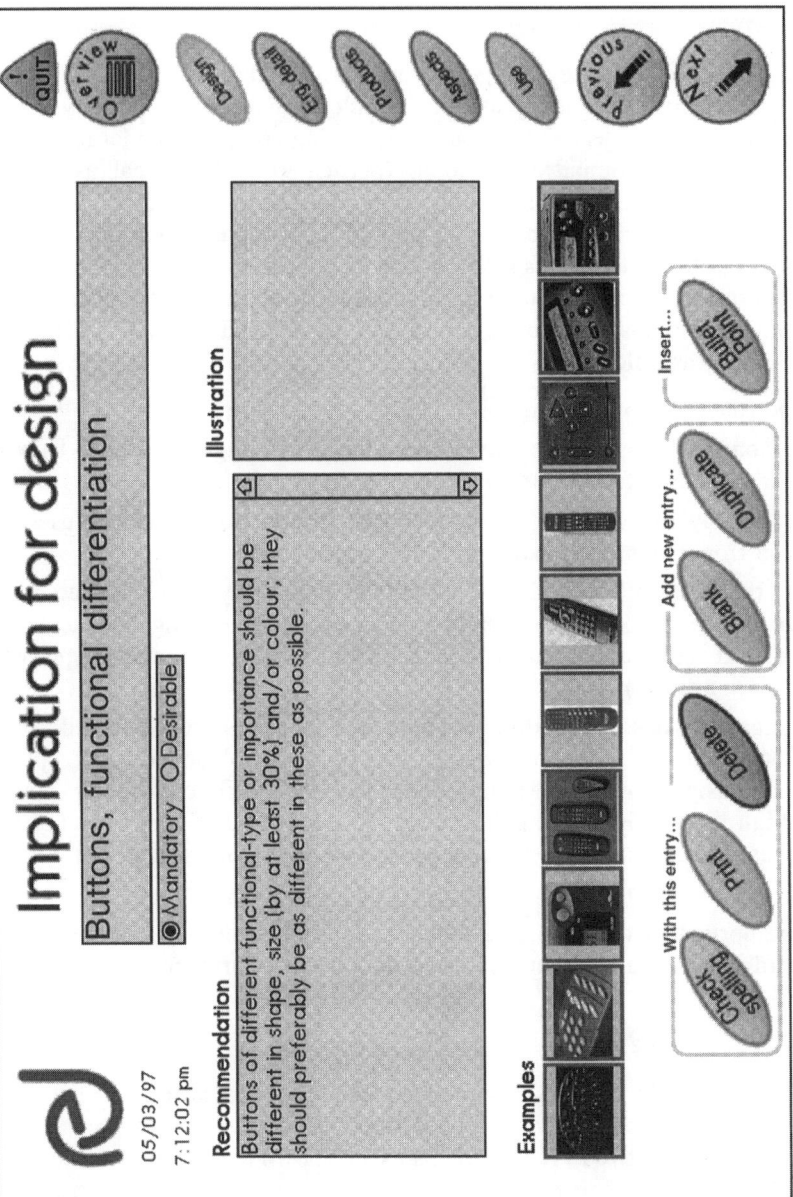

**Figure 6.2** Database of Recommendations for Ergonomic Design (RED).

features – by producing a number of different concepts through brainstorming and then evaluating these with consumers.

## Brainstorming

Brainstorming is a process of generating various ideas for how the product could be designed. For many aspects, simple sketches are a primary way of initially expressing ideas. These can evolve using computer renderings, foam/ wood mock-ups, computer modelling (particularly of physical aspects) or computer simulations of functionality as appropriate to the project. As well as being used as materials for consumer testing, these representations of concepts are useful as reference points for cross-functional discussion.

When the approach to be used by an interface is of interest, the configuration of controls and displays to be tested can be defined through a structured methodology as follows:

1.  The functions to be incorporated into the product are listed.
2.  A 'control' list is made of any information which might be needed from the user to the interface before performing each function.
3.  A 'display' list is made of any feedback information which the user may need from the interface after performing a function.
4.  To the 'control' and 'display' lists are added any possible methods of implementation (e.g. selection of a value can be done using discrete buttons for each value, up- and down-buttons with the value somehow displayed, a knob or a slider).
5.  Pictures of designs of different interface concepts are generated which, between them, cover the most interesting control and display options. Each design may differ from the others in many different aspects, both on a high level of interface philosophy and a low level of control and display types. These will be used to gauge the reactions of users to the different options.

It is important that each of the concepts presented already achieves a high standard of design and ergonomics according to the information known so far.

## Selection of a Single Concept

The first consumer testing of the new concepts is oriented towards defining a single solution which will be developed further.

Small groups of two to four participants are shown the different concepts and encouraged to state their preferences, first on a general level and then for each function or other aspect. It is important that the reasons for their preference are understood, and in doing this the discussion has to be flexible, like that of a focus group. It is unlikely that one concept is better in every way than the others – the results usually take various aspects from different concepts which might be later combined into a single design.

### 6.2.3 Concept Development

With the establishment of the elements of the single development concept, an iterative process of refinement and evaluation follows. Development of the dialogue through which a user and product communicate ('interaction design') requires special methods, and these are detailed separately in section 6.3.

*Initial Design and Later Refinement*

During development there are continual cross-functional discussions and refinements as the design matures and details are resolved. A concept that starts as drawings develops through a number of increasingly detailed and realistic physical mock-ups. Initial mock-ups of physical parts may be in foam-like materials and later wood may be used. Where only one aspect of a product is being redesigned, the illusion of reality is heightened by the rest of the product being real.

A balance must continually be struck between engineering constraints (including production cost), the 'purity' of the aesthetics and ease of use.

*Short-term Evaluation of the Single Concept*

The evaluations become increasingly detailed and long term. Throughout all of them the test results should be summarised and recorded for future reference (Whirlpool does this using its RED database mentioned earlier).

Laboratory-based evaluations of mock-ups are usually done using a combination of questionnaire-based rating scales for quantitative impressions (e.g. checking that aspects achieve the ergonomic standards set – see section 6.4.1) and open written or verbal questions. Care has to he taken to make sure that the user is focusing on the new design elements of interest without being sidetracked by the unrepresentative quality inherent in a mock-up (e.g. feeling of solidity and durability), or other details such as the kind of product that *they* would prefer.

*Tests of Functional Machines*

When the first prototype machines appear, the detail in the tests increases further with actual use of the physical machine being represented for the first time. Tests may be conducted with the consumers following the 'instructions for use' to evaluate both machine and instructions.

*Comparison against Competition*

The final laboratory-based evaluations are to benchmark the product against the those with which it was designed to compete in the marketplace (usually those benchmarked much earlier in the design process). This is intended to confirm that the concept has achieved its design objectives, but if it falls short there may still be time to make some changes – in any case, lessons can be learnt for next time.

*Medium- to Long-term Evaluation*

It is one thing for someone to come into contact with a new product for only an hour or less in a laboratory, but quite another to live with a product. The development process finishes, therefore, with two longer-term evaluations which are handled by the market research department: firstly the field testing of preproduction machines at the homes of representative consumers for a period of weeks or months, and secondly the polling of known users with postal questionnaires after a period of ownership. Whilst the former can have some effect on a design, even at this late stage in development, the information gained is likely to contribute more to the next generation of products.

## 6.3   Interaction Design

The concept development of the dialogue between the user and the controls/ displays of an interface – interaction design – is a relatively new area of work which is of increasing importance. The particular methods used for interaction design are described in this section.

The interaction development process generally begins with an initial definition of the design, followed by iterative cycles of evaluation and improvement.

### 6.3.1   Initial Definition of the Design

The design is built up in three general phases:

- Generation of fundamental principles and storyboards
- Detailed interaction design using moding charts and mode-feedback matrices
- Generation of an interactive simulation to test the interaction.

These phases are described here and illustrated using the example of an oven timer.

*Storyboards and Fundamental Rules*

The first phase of the design aims to define the fundamental ways in which the interface will work. This is done both by establishing a set of rules for the interaction and by using storyboards to give a basic representation of how the interface may appear.

Following the example of the oven timer, the underlying principles could be as follows:

■ When a value is being set it flashes on the display, otherwise (when it is current) it is steady.

■ The clock is normally shown unless the timer has been set, in which case the time remaining is shown.

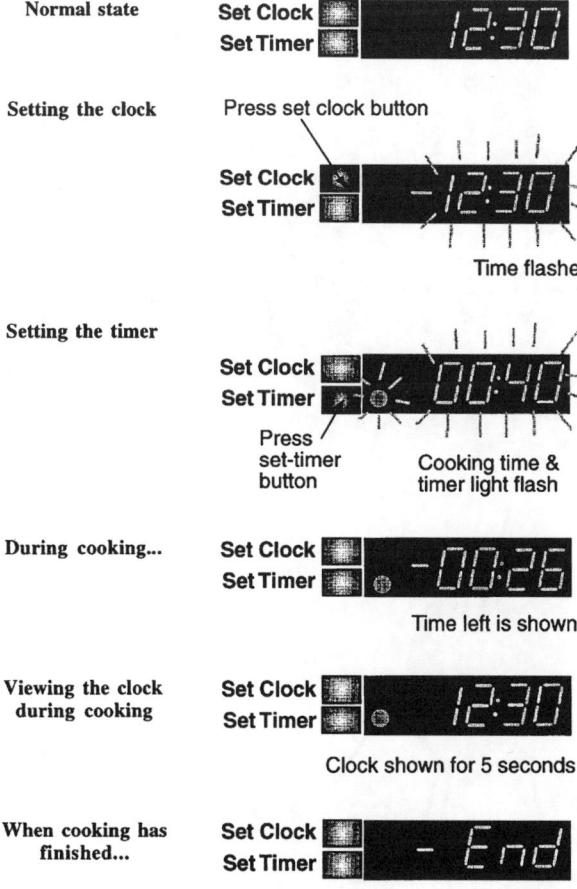

**Figure 6.3**   Storyboard for an oven timer.

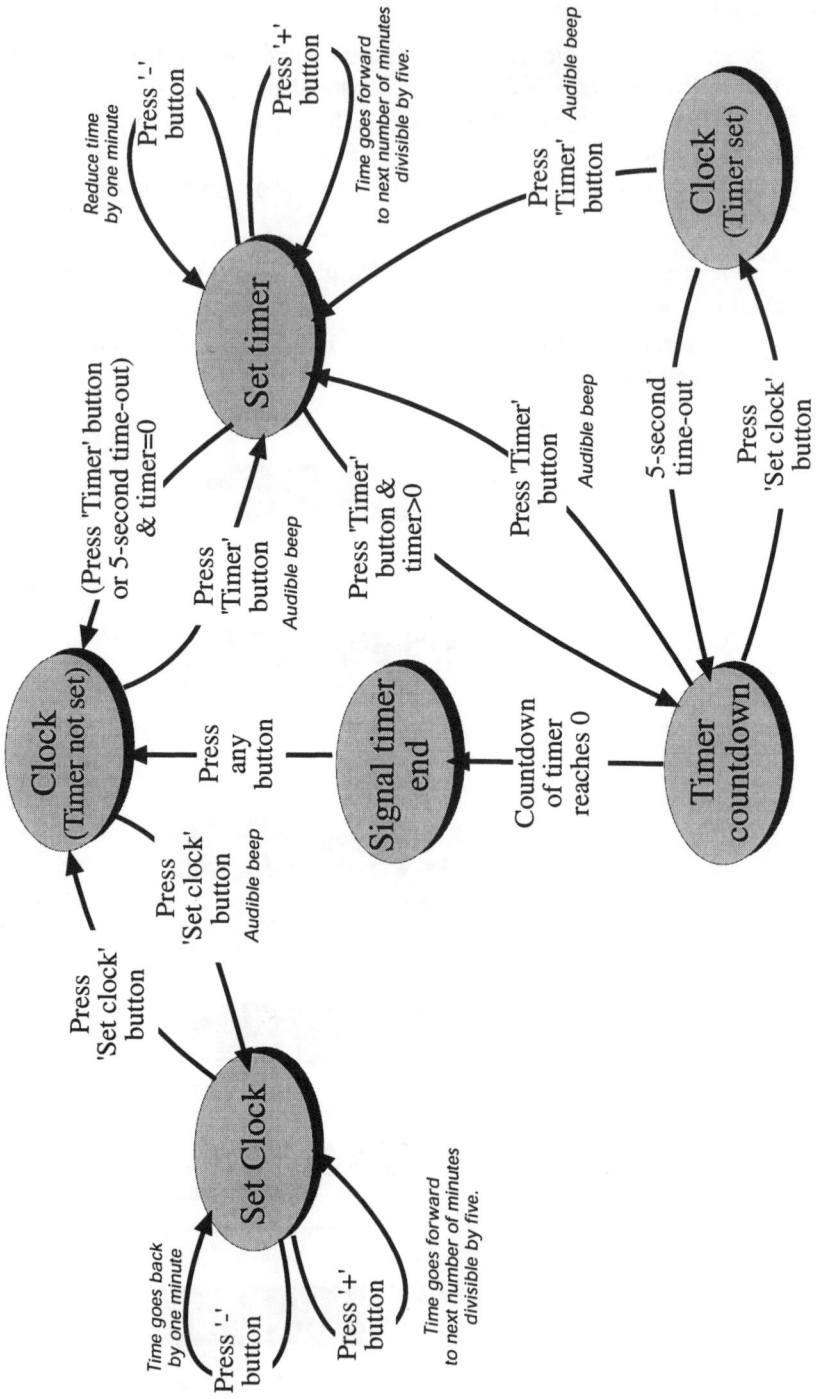

**Figure 6.4**    Moding chart of an oven timer.

■ If the user presses the clock button to set the clock, or the timer button to set the timer, then pressing this button again will return the user to the normal clock mode.

The storyboards typically show a series of pictures of the interface in different situations with a suggestion of how the mode was entered. A typical example is shown in Figure 6.3. Whilst the images may be drawn with a computer, sketches are also commonly used during the brainstorming process. It is important to emphasise that the storyboard is just an exploration of an interaction idea which becomes more 'solid' in the coming phases.

### Moding Chart and Mode–Feedback Matrix

The next part of the process is to add 'flesh' to the design – a detailed definition of the moding. This is done using a combination of a moding chart and a mode–feedback matrix.

The moding chart is a form of task analysis diagram used in defining the separate modes and the conditions for transition between them. The chart for the oven timer example can be seen in Figure 6.4.

Each of the modes is a stable state for the interface, with its own unique configuration of feedback given to the user (as specified in the mode–feedback matrix). Modes are shown in the chart as closed shapes – here ovals are used – with the mode name written inside.

Transitions between modes are usually caused by the use of a control or some other condition like a time-out (e.g. if the interface is in set timer mode and no time has been entered then the interface returns to the clock mode if no button is pressed within 5 seconds). These are marked as arrows; text over the arrow path indicates the condition for the transition, and additional text off to the side in italics gives extra detail about anything that happens during the transition.

The feedback given within each mode to the user, both visually and audibly, is defined using a mode–feedback matrix; that for the oven timer is shown in Table 6.1. One fundamental of interface design is that the feedback given for each mode should be unique and that it should follow certain guiding principles.

### Simulation

The final part of the definition process involves the production of an interactive simulation of the product interface. As well as being used in consumer evaluation of the design, it can also be distributed cross-functionally to facilitate detailed discussion of how features are accessed. Through the use of a multimedia development application called Macromedia Director, the simulations generated work both on the Macintosh computers used by the

**Table 6.1** Mode–feedback matrix for an oven timer

| Mode | Numeric display | Timer LED | Audible beep |
|---|---|---|---|
| | | Feedback | |
| Clock (timer not set) | Clock, steady | Off | – |
| Set clock | Clock, flashing | Off | – |
| Set timer | Cooking duration, flashing | Flashing | – |
| Timer countdown | Time left in format '- 0:00' | On and steady | – |
| Clock (timer set) | Clock, steady | Flashing | – |
| Signal timer end | '-End' | Off | Once a minute there is a sequence of 10 beeps (one a second) |

designers and on the PCs used by the rest of Whirlpool. They are normally distributed through electronic mail.

The designer usually produces an image of the interface which can be imported into Director as various elements (the panel background, buttons, knobs, displays and so on). These are then animated by adding instructions through Director's scripting language, Lingo.

### 6.3.2 Evaluation

The simulations are then used to look at how people access functions and form a mental model of the user interface. These studies may be particularly complex and the methodologies detailed by Rubin (1994) are used.

*Fidelity of the Simulation*

The ideal way to test interfaces would be to have a mock-up product with real controls and displays driven from a hidden computer. This would resolve most of the issues related to the artificiality of using virtual controls on a touchscreen, but creating something like this takes a great deal of time and resources. Consequently, usability testing at Whirlpool involves the simulation of the product interface on a touchscreen computer. This method creates many issues which have to be recognised:

■ Knobs or other features to be grasped and moved cannot be represented properly. Testing on the physical aspects of these is done separately using physical mock-ups, but when the control is an integral part of performing a control task under test then it has somehow to be represented on the screen. It is vital that the way that the user is intended to turn the knob is communicated and understood in order to prevent user difficulties in something which would not normally be an issue. The technique used at the moment is that rotation arrows are provided adjacent to the knob and users are told to click on them in order to turn the knob in a particular direction.

- Buttons depicted on the interface are also not as clearly defined as they would be on the real product and there can be an artificial ambiguity in communicating which features can be pressed. They therefore have to be emphasised graphically.

- The resolution of the touchscreen means that buttons can sometimes be difficult to activate. If the user has tried to 'press' a button and it did not respond, then the mediator may intervene to complete what the user intended. The problem with pressing buttons is especially acute when the buttons are small, and in such a case the sensitive area of the button may be made bigger than the button itself. A computer mouse has also been tried as a way of selecting buttons more positively, but users are often unfamiliar with its use and it makes metrics based on the time taken to achieve a goal unreliable.

The most recent 'physical' versions of the product are always on hand to put the simulation into perspective, and the presence of the mediator with the users makes sure that any artificial situations are compensated for as easily and as quickly as possible.

## Metrics and Results

Where the way in which the participant uses the controls in achieving a goal is important, then the user is asked to perform representative tasks. As well as the time taken to finish the tasks being recorded, any areas of confusion or error are noted.

Where situational awareness is an issue (i.e. understanding what the product is telling them), then numbered 'scenario buttons' on the screen are used to put the user into a situation which they are then asked to describe.

It is important to know as much as possible about the users' feelings while performing the tasks or reading the displays. Tests are video-recorded with a split screen showing the interface on one side and the faces of the participants on the other: this can show when the users are uncertain as well as what they are looking at or trying to press (their fingers are visible against the screen). The video also records any comments the users make, and the use of pairs of participants means that users are likely to vocalise any thoughts when talking to each other – this is more natural than asking a single user to think aloud. As video takes so much time to analyse, this method is used primarily to supplement observations and metrics recorded on paper. Examples of situations captured on video can be helpful in communicating issues at a subsequent results presentation.

### 6.3.3  Improvement

*Use of Functions*

The areas of confusion or error revealed during evaluation are applied back to the moding chart to reveal the improvements possible.

This can best be illustrated through a hypothetical example relating to use of the oven timer. The scenario is that the oven is in the process of cooking but that, despite being one minute from the end of the timer period, the food does not seem to be adequately cooked. The user therefore has to increase the timer period to three minutes left.

Figure 6.5 shows the difference between the intended and actual user actions in terms of the moding chart. With the initial mode being timer countdown, the solid transition arrows show the user actions intended by the designers:

- Press the timer button.
- Press the ' + ' button to select 5 minutes cooking time.
- Use the ' − ' button twice to reduce this to 3 minutes.

The dashed arrows show how the user actually attempted to fulfil the task:

- Press the ' + ' button twice to increase the cooking time from 1 to 3 minutes.

The mismatch between these shows that the user's mental model of how functions are accessed is different from that intended by the designer and communicated through the design of the interface. There are consequently two levels at which the design could be improved.

On a 'high level', the use of the controls should conform more closely to the user's natural instincts. In Figure 6.5 the user expected each press of the ' + ' button to advance the time by one minute. Whilst this is understandable, it would become inconsistent with the way the button has to work when selecting much larger times, and so the functionality of the button is left as it is (advancing to the nearest 5 minutes).

However, if most users feel that they should be able simply to press the ' + ' and ' − ' buttons to change the cooking time directly, then the design should allow this in a way that also accommodates the interface rules and overall logic and consistency of the interface. One possible improvement to the design could be that shown in Figure 6.6. The timer can still be changed in the original way:

- Press 'timer' to change mode such that the current setting flash.
- Change the setting with the ' + ' and ' − ' buttons.
- Press 'timer' again to finalise the setting and return to the 'countdown' mode.

Now, though, an additional way of doing the same thing can be added:

- The first press of ' − ' or ' + ' both changes to the 'set timer' mode and changes the setting up or down once (as appropriate).

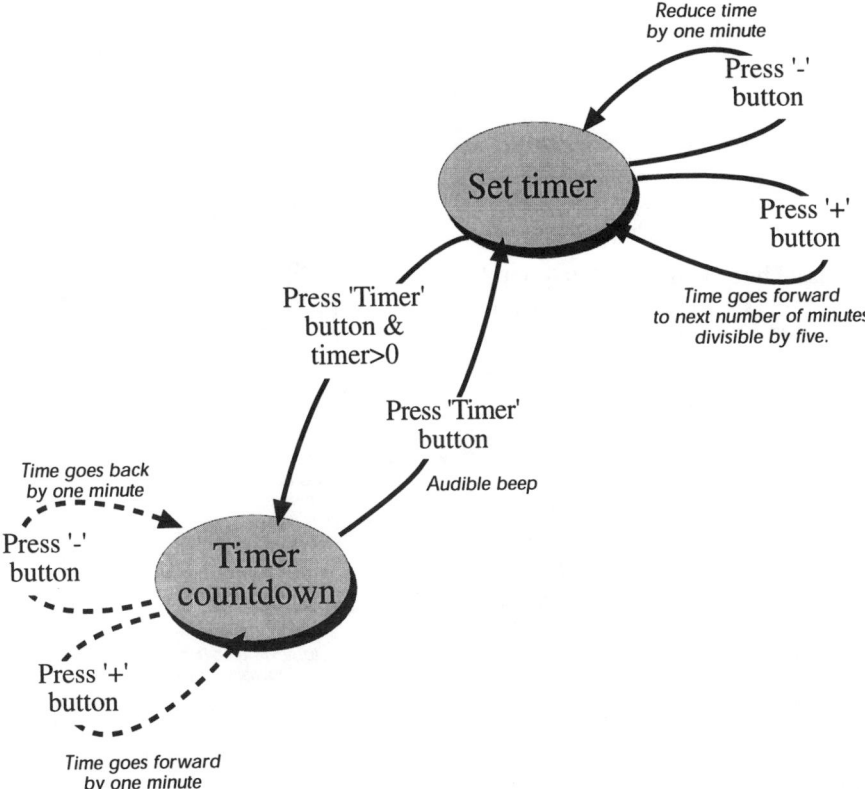

**Figure 6.5**   Actual and intended changing of timer setting.

- Subsequent presses of ' − ' or ' + ' continue changing the timer up or down.
- Not touching anything for a short time-out period (e.g. 5 seconds) returns to the original mode. This is important for 'closure' − the user does not directly make the mode change in the first place and so will not be expecting to have to change the mode back again.

It is good to have a flexible interface in which there is more than one way to achieve an effect as long as it is still consistent with the fundamental principles.

On a lower level, there is the issue of how an intended mental model is communicated to the user. This is a topic well covered by Norman (1988), who also uses the concept of 'affordances' which is useful here. Affordances are basically what a user thinks they can do with something (be it an object or an interface) in order to perform a task. In this context the affordances intended by the designer in accessing functions should be visible and communicated well through the use of 'design' in the buttons, display and panel, and in any

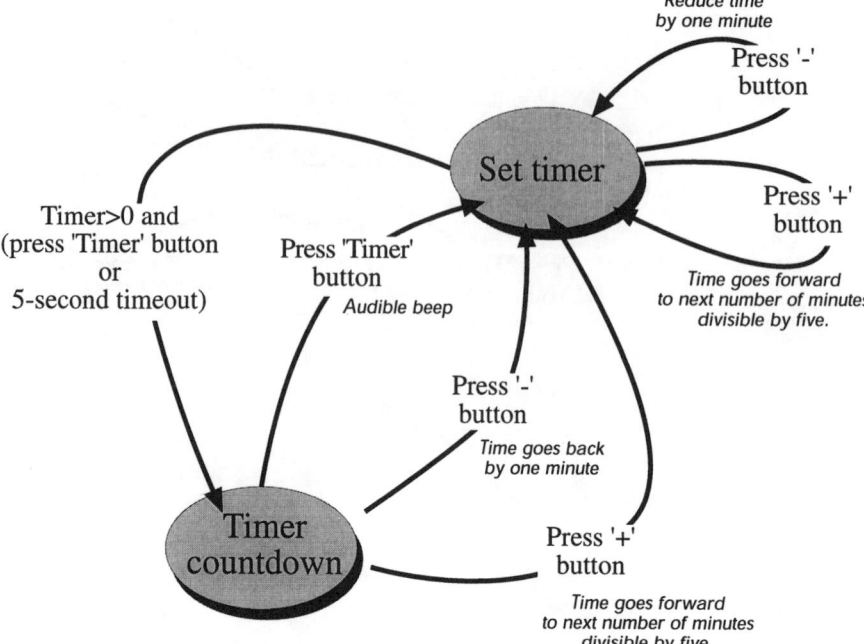

**Figure 6.6**    Possible improvement to changing of timer setting.

additional graphics. On the other hand, any affordances not intended should be hidden as far as possible so that they do not distract the user.

*Improving Communication of Information*

Communication is even more of an issue when users are put into a situation when using an interface and asked to describe it, or when they are asked to perform a task and misunderstand the mode that they are in. Any mismatch between what the user has understood and the real situation can studied using the mode–feedback matrix, and the appropriate displays or sounds highlighted.

## 6.4   Low-profile Ergonomic Development

### 6.4.1   Use of Standards

The development process described so far very much involves the ergonomist, but following this process for the many small parts of products, such as buttons, handles and knobs which are constantly being aesthetically rede-signed, would take too much of his or her time. Consequently a methodology is used for these parts where the ergonomist takes a lower profile and thus gains time to work on other issues.

It would be possible for the ergonomist to specify ergonomic guidelines to the designer with the attributes and acceptable ranges of dimensions for such parts, but this prescriptive approach can be considered by designers to be too intrusive and constrictive on the form and appearance. Consequently, in order to avoid forcing ergonomics on the designer from the outset, a standards-based approach is used (Feeney, 1996): this means that the designer has a high amount of freedom in the design of the part, but when the design is 'audited' with consumers it must achieve standards set for ergonomic, aesthetic and perceived quality.

Standards are defined in terms of parts of products and their various aspects (which may be related not just to ergonomics but also to perceived quality and aesthetics). For example, aspects for a push button might include:

- Size
- Comfort
- Apparent quality
- Appearance
- Response when pressed
- Pressing force needed.

### 6.4.2 Auditing

During a design audit, representative consumers are asked to rate the aspects for each part on a five-point scale of 'poor', 'below average', 'average', 'good' or 'very good'. Aspects which cannot be reliably represented by the products or mock-ups being shown will not be included. For example, a mock-up may not faithfully reproduce the feel of the buttons and so the participant will not be asked to rate 'response when pressed' or 'pressing force needed'.

In addition to the scales, users give open responses to any aspects which they particularly like or dislike.

### 6.4.3 Analysis of Ratings

Whenever a participant scores an aspect as either 'good' or 'very good' then this is considered a positive result. The standard is that at least 90 per cent of the ratings of an aspect should achieve this; the 'normal' way of determining this is illustrated on the left-hand side of Figure 6.7.

| Participant number | Rating (1 to 5) | Positive result (score 4 or 5) | | Demographically important | Weighted result |
|---|---|---|---|---|---|
| 1 | 2 | ✖ | | ✔ | ✖ ✖ |
| 2 | 4 | ✔ | | | ✔ |
| 3 | 5 | ✔ | | | ✔ |
| 4 | 4 | ✔ | → | ✔ | ✔ ✔ |
| 5 | 3 | ✖ | | ✔ | ✖ ✖ |
| 6 | 4 | ✔ | | | ✔ |
| 7 | 2 | ✖ | | | ✖ |
| 8 | 5 | ✔ | | ✔ | ✔ ✔ |
| | | 5 positives out of 8 62.5% | | | 7 positives out of 12 58% |

**Figure 6.7**  Weighted analysis of audit ratings.

However, the composition of the consumer groups would ideally be demographically 'tailored' according to the type of aspect under test:

- For more ergonomic aspects, the ratings given by older or disabled participants are more important than those from younger, able-bodied people. This is a way of linking the ergonomic standards with the concept of 'design for all' in which satisfying the more stringent needs of the elderly and disabled makes a design more comfortable for everyone.

- For aspects which predominantly relate to aesthetics, those participants who can be categorised as the target consumers for that product and brand are particularly important.

- Equal importance is given to all participants when the aspect is related to an impression of quality.

The constraints of testing mean that the same group of participants has to be used to test all these different aspects at the same time. If the ratings of the participants who are demographically 'less important' for a particular aspect were simply excluded in analysis then the final sample sizes may be too small and too biased towards the 'important' participants.

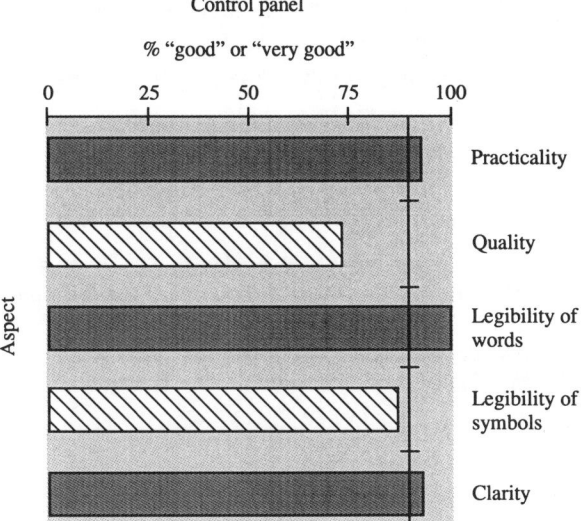

**Figure 6.8**    Consumer judgement of control panel aspects against standards.

Instead, the results for an aspect are biased in favour of the 'important' participants (e.g. an older person in a purely ergonomic test) by giving their ratings a greater weighting than the others. The best way of achieving this weighting is still under development, but it is currently done by simply duplicating the 'important' results – as if each had an additional twin – before again determining a percentage of positive results. This is illustrated on the right-hand side of Figure 6.7.

### 6.4.4    Representing and Acting on Results

Results are then communicated back to the designer. The example in Figure 6.8 shows the percentage of users giving positive ratings of aspects of a control panel – practicality, legibility of words and clarity meet the standard, whilst quality in particular needs attention.

If an aspect falls short of the standard then the designer and ergonomist will work together on improving the part in this respect. The open comments recorded during the audit can be a useful source of information on how this should be done, as well as other ergonomic guidelines.

## 6.5    Conclusions

Having described the methods by which ergonomics is included in product development in Whirlpool, it is important to point out that the methods are always undergoing improvements and that there are almost certainly others in use elsewhere which are as good or superior. Even so, the methods currently in use benefit the products greatly, as we hope consumers will discover in the coming years.

## References

NORMAN, D. (1988) The psychology of everyday things, *First person: Donald A. Norman. Defending Human Attributes in the Age of the Machine.* Voyager CD-ROM.
RUBIN, J. (1994) *Handbook of Usability Testing.* Chichester: Wiley.
FEENEY, R. J. (1996) Personal communication, Robert Feeney Associates.

# CHAPTER SEVEN

# Consumer products – more by accident than design?

MAGDALEN PAGE

*ICE Ergonomics, Loughborough*

## 7.1 Introduction

The evaluation of consumer products by ICE Ergonomics goes back to the origins of the organisation itself in 1970. Indeed, those with long memories will remember that ICE stands for the Institute for Consumer Ergonomics, thus indicating the emphasis of its early work. Over the intervening years the balance has shifted away from consumer products, but this area remains an important part of ICE Ergonomics' work.

Publications by senior staff involved with ICE Ergonomics on the subject of the methodological aspects of the topic also go back to 1970 (Kirk and Ridgway, 1970, 1971; Wilson and Kirk, 1980; Rennie, 1981; Fulton and Feeney; 1983).

It is also important to note that although this chapter is concerned with the critical evaluation of existing products, similar techniques can be applied to the constructive evaluation of the design of a product under development. The main difference is the timing of the ergonomics input and the willingness on the part of the designer or manufacturer to adopt and incorporate the ergonomics recommendations into the design. Self-evidently, if one is evaluating a product that has been involved in an accident, the manufacturer of which is about to be sued by an injured consumer, then the comments come too late to help. Conversely, if the input is available whilst the product is still on the drawing-board, then, at minimum cost, the recommendations can be considered in a positive light. Table 7.1 indicates the contribution that ergonomics can make at each of the stages in a product's lifecycle.

The techniques and methodologies described are, where possible, illustrated by examples; however there will be times where confidentiality has to be

127

**Table 7.1** The role and contributions of ergonomics in the design process

| Design process stage | Ergonomics input |
| --- | --- |
| Initial concept or idea of the function of the product | ● Data on:<br>  – human requirements<br>  – human needs<br>  – human shortcomings |
| Functional specification | ● Data on:<br>  – human capabilities<br>  – human size<br>  – human strength<br>  – accidents |
| Design drawings and pre-production prototypes | ● Evaluations or appraisals<br>● Recommended changes and modifications<br>● Identification of potential hazards and mismatches |
| Production and marketed products | ● Evaluations and appraisals<br>● Anticipation of foreseeable misuse<br>● Instructions and warnings |
| Accidents and complaints | ● Accident investigations<br>● Interactions with the environment and other products<br>● Misuse |

maintained. There is also a brief mention of the role of accident data analyses in the process, as well as a pointer to the legislation that supports and underpins the evaluation process.

The intention is to give some insight into the practical ways in which ergonomics contributes to the design and the evaluation of products at ICE Ergonomics. ICE Ergonomics uses an essentially pragmatic approach that is mindful of the timescales and budgetary constraints of its clients. However, ICE endeavours to maintain scientific rigour and validity in its approaches.

## 7.2 Home Accidents and their Contribution to Product Evaluation

### 7.2.1 Introduction

An accident may cost the victim not only pain and inconvenience but also considerable sums of money. Furthermore, the interests of the manufacturer of any product involved may be seriously harmed if they become discredited when their products are labelled as unsafe. The financial consequences of a product recall, a criminal prosecution by Trading Standards Officers or a successful product liability claim may also be considerable. On these financial bases alone it is therefore vital that product evaluations are carried out to reduce the frequency and severity of accidents involving products.

Typically, during the evaluation of any given consumer product, one or all of the following stages are involved:

- First identify a product or group of products as constituting a risk.

- Look in detail at the nature of the hazard(s) by investigating accidents.

- Begin to look forward at current versions of the products rather than back at old products.

- On the basis of the information collected, develop criteria for improving the safety of future designs.

- Convey this information to manufacturers, designers and standards organisations so that recommendations can be implemented.

This approach is applicable to virtually any consumer product, since, by examining accidents, we can identify activities and product features that are critical to safety as well as where there are mismatches with users that have resulted in an injury. Current models and prototypes can then be examined in a structured way, and features of poor or hazardous design can be highlighted before they are actually involved in accidents.

Thus by studying accidents and subsequently using the information to improve the ergonomic design of products it is possible to:

- take account of the way people really use products;
- reduce the frequency and severity of accidents;
- make significant interventions in the safety of future products.

This makes it possible to define criteria for design in terms of human performance and to decide upon priorities which will result in a generation of safer products.

### 7.2.2   Accident Data Sources

It is clear that a full understanding of the use of any consumer product requires investigation of its involvement in accidents. Whilst it may not be possible to discover accidents in which a specific brand of product is implicated, it is enlightening nevertheless to know the numbers and kinds of accidents with which products of that general type are associated. The most accessible and fruitful source of accident data is the annual report published by the Consumer Safety Unit of the Department of Trade and Industry. In this document the interested reader can find out about the accident frequency of a whole range of individual products and environmental features, as well as see tabulated statistics presenting a wider view of the home and leisure accident situation in the UK. For those requiring a greater breakdown of the accident statistics, the Consumer Safety Unit will provide detailed computer printouts.

An alternative source of data is to contact specialist organisations who collate such accident data. These include the Royal Society for the Prevention of Accidents (RoSPA), the Child Accident Prevention Trust (CAPT) and the Health and Safety Executive (HSE). The names of addresses of these organisations are included at the end of the chapter. All three of these collect specialist information, including press cuttings, and can often assist in identifying problems with products that have been involved in accidents. These organisations also have specialist libraries that may contain relevant publications and information.

### 7.2.3   The Facts

In the UK, accidents in the home account for large numbers of visits to the accident and emergency departments of hospitals, and in 1990 they accounted for one-quarter of all male and almost half of all female accidental deaths (Department of Health, 1993). In the UK there are more deaths in the domestic environment than occur on our roads, and in 1994 there was a total of 130 992 non-fatal home accidents recorded in the UK's Home Accident Surveillance System (DTI, 1996), approximately half of which involved consumer products of some type. This is a small rise on the figures for 1992 and 1993 and provides a national annual estimate for all home accidents in 1994 of 2 658 000.

Whilst not all of these accidents can be prevented and the consequential injury avoided, good ergonomic product design can make a major contribution to reducing their numbers.

### 7.2.4   Accident Investigations

When one wishes to find out more about the relevance of a product's design or characteristics to an accident and to investigate real, as opposed to assumed, behaviour, it is often useful to carry out a detailed analysis of a selection of accidents. However, this can be complex and expensive, not least because special arrangements have to be made to contact the individual accident victims via the hospital that they attended to receive first aid treatment. Nevertheless, in the case of a major study it is worth the effort as this analysis of accidents allows identification of patterns of behaviour that, coupled with a particular product, led to an accident. Knowing how a product was being used allows a fuller understanding of the problem and highlights likely preventive measures which may be taken to guard against a repetition of the accident.

What frequently transpires is that most people who have accidents do not consciously take risks on the basis of a logical decision. In most cases they report that they had no perception of there being a risk at all and they were frequently familiar with the product in question. This leads us to conclude that the oft-quoted idea of reducing accidents by increasing the awareness of the user is somewhat pointless in the case of many familiar products (Weegels, 1996).

*Methodologies*

There are several ways of following up accidents and each has advantages and disadvantages. The methods most commonly used in studies at ICE and elsewhere (Weegels, 1996) include:

- surveys using questionnaires or telephone interviewers;
- accident reconstructions and home visits to victims.

The use of postal questionnaires can be a relatively inexpensive way of carrying out accident follow-ups since it does not require researchers to travel all over the country visiting people in their own homes. However, it does require considerable effort to be put into the development of the questionnaire, since the accident victims will usually be completing it without any support from the research team. The questionnaire needs to stand alone and be as quick and as simple to complete as possible, which usually means using multiple choice questions and simple tick-boxes. However, this presupposes that the range of likely answers is already known and precoded into the questionnaire. Additional research into the nature of the accidents and piloting the question- naire with a small number of home visits is therefore essential to ensure that the questionnaire will work when it is sent to the general public. In using postal questionnaires one must also bear in mind that the response rate for unsolicited questionnaires can be extremely low.

Telephone interviews have the advantage of being more interactive than a written questionnaire but have the problem that diagrams cannot be used easily, which can be a problem where complex products such as power tools are involved in accidents. ICE has surmounted this problem by sending advance information to accident victims which has the double advantage of giving them prior notice of our telephone call and allows us to 'talk them through' a diagram without the need for technical terms. The disadvantages of telephone interviews are that they often have to be conducted in the evening and at weekends, not everybody has a telephone, and many people do not notify the hospital of their number and so it is difficult to obtain.

Accident reconstructions during home visits are particularly useful as they allow subjects to demonstrate what happened whilst the researcher records the detail. It can sometimes be easier for the accident victim to overcome their reticence to describe what they may feel was their own stupidity if they can demonstrate it two or three times rather than having to express it in words. Although home visits can prove costly and time consuming, the quality of the data is generally very high and they can often save time at later stages in the study. A useful and comprehensive checklist for conducting in-depth accident analyses is given in Weegels (1996). It covers:

- the use of the product;
- the details of the product;
- social elements;
- the subject's details;
- the environment in which the accident occurred;
- a measurement of the forces that the subject can exert;
- an inspection of the product if possible.

It also suggests ways of recording the information for future analysis.

### 7.2.5  Identification of Hazards

On the basis of the study of accidents it can be helpful to carry out an analysis of the hazards* responsible for the accidents and injuries. The sequence of events leading to an accident have been described by Ramsey (cited in Hoefnagels and van Aken, 1991) and it is evident that the chance of being involved in an accident depends on a number of factors:

- exposure to a hazardous product;
- perception of the hazard;
- cognition of the hazard;
- decision to avoid the hazard;
- ability to avoid the hazard.

In a recent study of large self-assembly items for children (Page *et al.*, 1996) it was necessary to identify the scope and relevance of any hazard that was presented by the selected items, in order to assess the consequences that any misassembly may have had. From the accident data and other sources of information, it was clear that the hazards were wide-ranging and could affect the person assembling the item, the user or a third party. Similarly, accidents could occur during assembly or during use. From the accident data and other

---

* A hazard is defined as a potential source of harm, and harm relates to physical injury and/or damage to health or property. These definitions are taken from ISO Guide 51 (1990) *Guidelines for the Inclusion of Safety Aspects in Standards*, cited in Hoefnagels and van Aken (1991).

available literature, the main types of hazard that the items under investigation presented were established. These included:

- entrapment of the body or clothing in fixed or moving parts of equipment;
- cuts, lacerations or splinters from sharp edges;
- foreign bodies and particles in the eyes;
- falls from the equipment;
- instability of the item, or part, causing it to fall;
- part of the equipment breaking or the equipment collapsing;
- loose fixings resulting in ingestion or causing collapse of the item;
- protrusions on which clothing or parts of the body could catch or become entangled;
- DIY-related accidents.

Accidents relating to each of these hazards could be found in the accident data and related to both the product design and the assembly process.

## 7.3  Legal Framework

### 7.3.1  Introduction

To ensure their own viability and the safety of consumers, producers must ensure that their products are safe, efficient, comfortable and satisfying to use, as well as durable, serviceable and realistically priced. Thus, safe and ergonomic design of products must be seen as a positive business activity rather than as an additional production cost (Wilson, 1983, 1984). However, we must be aware that no product will be absolutely safe since this remains unattainable and would be an unbearable cost to industry and put innovation at risk. What is required is a reasonable level of safety that safeguards the consumer whilst still allowing manufacturers to produce products at an affordable price. Three major pieces of UK legislation that provide the legal framework for consumer safety are summarised below.

### 7.3.2  Product Liability and the Consumer Protection Act 1987

The basis of strict product liability is that any person suffering injury or damage because of a defective product can obtain redress, usually from the manufacturer of that product, regardless of whether the latter was negligent in allowing the defect to exist. In the UK the Consumer Protection Act 1987 sets out product liability provisions which came into force on 1 March 1988. Any product supplied after this date will be covered by the Act and a product will fail to comply if it is shown to be defective. Defective in this context does not mean faulty; rather it is one in which the safety is not such as people generally are entitled to expect. It attempts to be objective by considering both consumers and producers in general. However, a product is not defective just because safer ones are subsequently marketed or because it is of poor quality. Other laws cover this latter aspect of a product. When considering whether or not a product is defective the court will take into account all the relevant circumstances, including:

- how the product is marketed, that is the expectations that the advertising might raise, who the product seems to be aimed at and any uses of the product that the marketing depicts;

- any instructions or warnings that are given with the product, including anything that is left unsaid or badly expressed, the size of print and so on;

- what might reasonably be expected to be done with the product; this includes the sorts of things that users often do with products that may or may not be the primary use intended by the manufacturer, such as opening tins of paint with screwdrivers and standing on kitchen stools;

- the time when the producer supplied the product, to take account of our developing knowledge of safety issues over time.

It is easy to appreciate that an ergonomist can have a vital role to play in evaluating a product to see if it complies with these requirements. Ergonomics can be used to evaluate text, to determine reasonable use of products and to consider how the nature of an advertising campaign for a product might influence a user's expectations of the product. Product liability is an extensive area of specialist ergonomics evaluation that cannot be dealt with in detail here, but one of which all ergonomists need to be aware. Interested readers will find many books and journals on the subject of product liability.

The second part of the Act deals with the criminal aspects of supplying unsafe products and has now largely been superseded by the General Product Safety Regulations 1994, details of which are given below.

### 7.3.3   General Product Safety Regulations 1994

The General Product Safety Regulations 1994 came into force on 3 October 1994 and they follow closely the general safety requirement in section 10 of the Consumer Protection Act 1987. The purpose of the Regulations is to ensure that all consumer products are safe as defined in the Regulations and to lay down a framework for assessing that safety.

In the Regulations a safe product is defined as:

> any product which under normal or reasonably foreseeable conditions of use, including duration, presents no risk or only the minimum risk compatible with the product's use and which is consistent with a high level of protection for consumers.

The safety of a product will be assessed by considering a number of matters, in a similar way to the Consumer Protection Act, namely:

- the product's characteristics;
- its packaging;
- instructions for assembly and maintenance, use and disposal;
- the effect on other products with which it might be used;
- labelling and other information provided for the consumer;
- the categories of consumers at serious risk when using the product, particularly children.

Again, all of these are useful pointers for ergonomists when carrying out an evaluation of a consumer product, whether or not it has been involved in an accident.

When assessing the safety of products, the General Product Safety Regulations lay down a hierarchy of rules and regulations that can be used, and these include European product directives, UK national law, European standards, British standards, industry codes of good practice and, finally, the state-of-the-art technology. In some cases the relevant standards include ergonomics performance criteria that may be used in drawing up a checklist for carrying out an expert appraisal of a product (see section 7.4.4 below).

### 7.3.4  Sale and Supply of Goods Act 1994

This Act amends the older Sale of Goods Act 1979 and is primarily concerned not with the safety of goods but with the rather nebulous concepts of quality and meeting buyers' expectations. In determining quality the following factors are taken into account:

- fitness for all the purposes for which goods of the kind in question are commonly supplied;
- appearance and finish;
- freedom from minor defects;
- safety;
- durability.

Again, all of these are helpful to ergonomists in determining whether or not a product is satisfactory and safe for the public to use.

## 7.4 Methods of Ergonomics Evaluation

### 7.4.1 Introduction

Various methodologies have been used at ICE Ergonomics to carry out ergonomics evaluations, but at the heart of all of them has been the need to involve the public in product testing. Typical methods employed are:

- observation of real life use of products;
- discussion or focus groups with users of the products in question;
- user trials with real subjects.

In some cases it is necessary to carry out expert appraisals either instead of, or as well as, these tests with users. Such approaches to consumer product safety and design have been used successfully with a range of different products in research at ICE Ergonomics. These products have included lawnmowers, equipment for elderly people, power DIY and gardening tools, ladders, bicycles, child equipment, fitness equipment, and large self-assembly items for children.

### 7.4.2 Real-life Observations

Observation can be an essential precursor to any evaluation as it can enable the ergonomist to decide on the appropriate user tests for the product in question. It can be particularly helpful in identifying product misuse which will often be of greater importance from a safety point of view than the proper use for which the product was intended. Observation also allows the ergonomist to perform a task analysis which involves listing systematically all of the operations involved in the full use of the product. Once this is done the relevant general performance criteria such as safety and comfort should then be associated with each task.

*Overt observations*

Overt observations have the advantage of being easy to set up since the person being observed has agreed to the observation and is co-operating. The disadvantage may be that what is being observed is not reality but some edited version put on for the benefit of the observers. However, in many cases the information obtained can be an essential pilot stage to a larger-scale user trial since only by seeing the sorts of problem that ordinary people experience with a product can a proper set of trials with appropriate test criteria be established. ICE has used this technique to study behaviour using ladders where the effects of the user's personal circumstances and domestic arrangements have a direct bearing on the purpose and method of use of the ladder.

A special subtype of overt observations includes home placement of products with users for a period of time. In these cases users might have the product for 1 or 2 weeks so that a whole cycle of use can be studied. They might be asked to keep a diary of use during the time and to report on any problems, or be given a series of tasks to do and report on. They will then be observed using the product at the end of the period when they will be more familiar with it. Such techniques have been used successfully for a variety of baby products such as baby baths and feeding bottles, where the interaction with the baby cannot be assessed in the short duration of a user trial in the laboratory.

*Covert Observations*

Covert observations fall into two main categories:

- observation of the general public when they are largely unaware of the observation;
- observation by means of video cameras or through one-way mirrors where the users know they are being watched but the observer's presence is unobservable and therefore more easily forgotten.

ICE has used the former technique when watching users on shop escalators to see how they behave in a given circumstance; the latter technique is particularly useful where the behaviour of unsupervised children is of interest. This method was used for testing non-reclosable drug packs when it was required to see what the children did with the packs when they thought that nobody was watching.

Covert observation may well give a more realistic view on product use but there are ethical problems with watching people without their consent. However, in a public setting where surveillance cameras are the norm, using the video for the collection of behavioural data is generally justified in the interests of public safety.

### 7.4.3 Discussion or Focus Groups

The use of carefully constituted discussion or focus groups can provide extremely valuable information about the real-life use of products. It is also a way of obtaining the views of typical users of products regarding additional functions and design changes.

Most groups are made up of users or potential users of the product under test, and the optimum size of a group is around 10 people; any more can be unmanageable and fewer can lead to the domination of the group by forceful individuals.

In order to function well the group needs to have a carefully structured agenda which covers the topics under discussion. Figure 7.1 gives the agenda for a group that was discussing the need for harnesses on children's products.

The discussion must be carefully guided by the researchers and any points of

---

1  Introduction                                                   *(3 mins)*
   Give information on the function and purpose of the meeting

2  Background                                                     *(10 mins)*
   - What items of nursery equipment do you use?
   - Where, and how frequently, do you use them?
   - Which of these items, if any, do you use with some form of harness?
   - Attitudes to child safety and accident prevention.

3  Expectations                                                   *(15 mins)*
   - What ages of children would use each of the products?
   - What reasons might you have for using/not using a harness?
   - What features are you looking for in each of the products?
   For each product would you expect to
   - Not need to use a harness?
   - Use your own harness attached to the product?
   - Use a harness supplied as part of the product?
   - Use a harness supplied as part of the product but with the additional use of your own harness?
   Do you assume that the product is safe as it comes and that you should not have to worry about whether it is safe or do you take any further safety precautions?

4  Children's Behaviour                                           *(5 mins)*
   How does your child (children) behave whilst in a highchair, pushchair or baby carrier?

5  Previous Experience                                            *(30 mins)*
   Please describe fully any incidents, accidents or near misses involving highchairs, pushchairs or baby carriers and for each incident record details.
   What types of harness have you used, which do you prefer and why?
   What types of buckle mechanism have you used, which do you prefer and why?

6  Problems and Concerns                                          *(5 mins)*
   - What problems have you had using (or not using) harnesses?
   - Do you have any other concerns about the use of harnesses?

7  Expectations                                                   *(3 mins)*
   - Having spent the last hour and a quarter thinking about harnesses and pushchairs, highchairs and baby carriers, have your expectations changed in relation to them?
   - How would you feel if the nursery products you have been discussing did not have a harness or attachment points fitted as standard?

8  Summary and Thanks                                             *(2 mins)*

---

**Figure 7.1**  The agenda for a discussion group on harnesses on children's products.

note recorded on a flipchart or similar device for the benefit of the participants. It is essential to keep the discussion focused on the topic in hand as it can easily stray into areas of general interest. In the case of the harness discussion groups, it was particularly important to keep participants carefully focused as four groups were conducted in the UK and matching groups were also drawn together with a similar agenda in France, Sweden, Norway and Portugal in order to gain a European view on the attitudes and perceptions of parents to the provision and use of harnesses.

### 7.4.4  Expert Appraisals

An expert appraisal is independent of users and requires an expert to evaluate the product using his or her acquired experience and knowledge. Appraisals can be conducted where user trials would be too expensive and time consuming or when the product under test might be too dangerous for testing by the general public, e.g. chain saws. However, one must bear in mind that there have been studies reported in which the expert's judgement was shown not to be valid (Shackel *et al.*, 1969). In the case of choosing an expert for an appraisal, this can be either an ergonomist or someone who is knowledgeable about the particular product.

*Ergonomists as Experts*

Ergonomists tend to carry out expert appraisals by means of specially developed checklists (Figure 7.2). Such checklists can ensure that all aspects of the product and its use are considered and can aid reliability when more than one ergonomist is involved.

Expert appraisals are particularly useful for carrying out investigations into products that have been involved in accidents or where there are serious doubts about their safety. In such cases the ergonomist can consider the abilities of a range of consumers and then go on to act as an expert witness, preparing a report and, if necessary, appearing in court. Expert appraisal can also be used as a precursor to user trials. In the case of a study at ICE into the evaluation of large, self-assembly items the initial expert ergonomics appraisal considered:

- the packaging;
- the ease of identification of component parts;
- ease of assembly;
- tools required or provided;
- problems during assembly;
- aspects of the design that caused concern;
- presence of assembly instructions;
- the age range for which the product is intended;

---

**Checklist for evaluating safety gate**

- Is the gate resistant to push and pull forces generated by a child?
- Is the gate free from protrusions, points, edges, burrs, flashes, or splinters?
- Are any gaps large enough to prevent finger entrapment but small enough to prevent hand insertion and body part entrapment at all times through the ranges or movement of the latch and hinges?
- Is there sufficient gap between adjacent horizontal members to prevent potential footholds? If there are any features, such as hinges, catches, decorations and fittings, are they narrow enough not to be an adequate step?
- Is the vertical distance between the highest attainable foothold and the top of the barrier sufficient to prevent a child climbing over?
- Does the latch require at least 2 distinct actions to operate it, one of which is an upwards force?
- Is it easy to tell when the gate is correctly closed?
- Is it possible to open and close the gate with one hand?
- When the gate is open, is the gap wide enough for an adult (possibly a pregnant woman) to pass through easily?
- Is any bar that must be stepped over to go through the gate as low, narrow and rounded as possible, to prevent tripping?
- Does the latching mechanism self-latch?
- Do the instructions include all the necessary information, such as safe assembly procedures, installation, removal, use and maintenance, as well as warnings against potential misuse?
- Can the gate be installed or removed without tools or technical expertise?
- Are all materials resistant to a child's bite, UV, etc. and will not degrade over time?
- Will the gate remain safe and usable for all of its expected lifespan, including when sold as second-hand?

---

**Figure 7.2**   An ergonomics checklist for evaluating a safety gate.

- presence of other information accompanying the product;
- the time taken by the ergonomist to assemble the product.

The ergonomist can also carry out an expert appraisal of the instructions that accompany a product where the following aspects will be borne in mind:

- *Physical description* of the instructions.
- *Language*   Is the language used easily understood by those who will assemble the product? What about special user groups, e.g. people for whom English is not their first language, or those who have difficulty reading?
- *Content*   Are the instructions comprehensive and is all the necessary information included?
- *Typography*   Are the typeface, print size and spacing appropriate?
- *Warnings*   Are all the necessary warnings and safety messages included?
- *Format*   Are the instructions of a suitable form and durability to remain usable for as long as they are needed?
- *Organisation and layout*   Is everything sensibly organised/grouped and properly laid out?
- *Usable*   Are the instructions printed in a way that is easy to read in the circumstances in which the equipment is likely to be used?
- *Illustrations*   Are clear and appropriate illustrations included where necessary?

*Product Experts*

Product experts can give a broad perspective when appraising a product, but there is a danger that they might be biased and may overlook a crucial point highlighted by inexperienced users. Consequently it is possible that if the results of expert appraisals and user trials are compared, they will not always agree. However, this is not a problem confined to expert appraisals as it is not unusual for user trials to provide contradictory results. In one study of the performance of door locks, the best lock, when judged on the basis of the objective results (the time taken to undo a lock in various environmental conditions and the number of errors made), was not the one preferred by the subjects who made their choice on the basis of aesthetic appeal and appearance (internal ICE client report).

### 7.4.5   User Trials

User trials are what most people think of when considering ergonomics evaluations. They are the most valuable source of information about a product's performance and can provide the best quality of data on which to make a decision to change a design or make a new product.

*Subject Selection*

User trials typically involve a panel of subjects drawn from a population most likely to use or have used the product. In some cases, market research data will be needed to identify who these people are; in other cases the client will provide details of his customer base. Once the population has been identified, the size of the sample must be specified – a topic that always invites hot debate since there is usually a balance to be struck between what is statistically and scientifically significant and what is economically justified. Today it is rare for a product to be tested by means of full-scale user trials in which a large sample of statistically representative consumers systematically tests a product under laboratory conditions. It is now more typical to use a sample of perhaps 20 people, who represent a typical range of users and who then test the product according to a carefully prescribed programme of tasks.

A second consideration in subject selection is whether the users should be naive or experienced. With more technical products there is a tendency to choose those who have some experience, especially if there is any hazard associated with the product, as in the case of power tools. In such cases the experienced user will bring skills, training and knowledge to help him or her make a more informed evaluation of the product. However, such an experienced user may well be biased and will not highlight the problems faced by a person using the product for the first time. Conversely, however, the totally naive user may be so overwhelmed by the sight or feel of a strange product that they may contribute little or nothing to the evaluation.

What is needed is a balance of users with a good range of skills. However, it is worth noting that the ergonomist may well need to assign users to the 'experienced' and 'naive' categories after the testing since experience at ICE indicates that many test panellists overestimate their level of skill when asked to rate it themselves. Such problems can lead to difficulties in subject selection and representativeness.

In addition to experience with the product in question, it is often necessary to select, or at least categorise, the sample on the basis of other characteristics such as:

- age;
- sex;
- anthropometric dimensions relevant to the product, e.g. grip span;
- product ownership.

## Market Research

Before carrying out user trials on a range of similar products, as in comparative testing, it is useful to conduct a market survey to find out what product models are currently available for the public to buy. From this spectrum of potential products a subsample must be chosen that represents the whole range. In making this choice, such things as price and availability need to be borne in mind, especially if the results are to be extrapolated to all products on the market.

Once the selection has been made, the products have to be bought, ideally from conventional retail outlets to ensure that the items under test are representative of those available to consumers in general. The temptation to get samples directly from manufacturers needs to be avoided as there is always the concern that such examples will be special in some way.

## Development of the Test Protocols

Once the subject panel has been identified and the products for testing have been purchased, the next step is to define the test protocol itself. This will typically involve the following stages:

- Preparation of a *detailed task analysis* which may well flow from the initial expert appraisal already carried out. This involves preparing a list of all of the activities associated with the product in a logical order from initial unpacking right through to emptying of any waste produced and disposal of the item itself, if appropriate. Such a task analysis will lay down the various activities that the subject will be asked to perform and on which the product's performance will be judged.

- Specify the *test location* to facilitate the testing programme. There is a need to balance the value of carrying out research in a laboratory where the conditions are controlled but unfamiliar and unnatural, with the more realistic but uncontrolled circumstances of a home environment. Some organisations use test apartments in which the users can be observed through one-way mirrors whilst they carry out the various activities. However, such a facility is expensive.

- Once the details of the task have been specified there is often a need for *apparatus* to carry out the tests. This apparatus can include anything from measuring equipment (anthropometers or force platforms) to a complex test facility. As an example of specialist measuring equipment, ICE developed a rig with load cells for the express purpose of recording the dynamic loads imposed on ladders during their use. The rig, which consisted of two separate load cells, one of which was used beneath each of the stiles of the ladder, was capable of measuring simultaneously in the three orthogonal planes at a very high sampling rate. The output was passed directly into a computer for analysis by specially developed software. It was then possible to measure and analyse the patterns of the loads exerted on the ladders as a series of subjects ascended and descended them and carried out various typical ladder activities at the top. A less technical but equally specialist example of a test facility is one that ICE developed to evaluate the performance of harnesses in children's pushchairs. An 'assault course' was devised that would put the selected pushchairs through their paces in a realistic but controlled environment. This involved a variety or terrains (gravel, uneven paving slabs, etc.), and obstacles (steps and slopes). Such apparatus needs to be specified, built and then piloted to ensure that it is performing correctly.

- Once the exact details of the trial are known, the various *data collection methods* to be used can be specified. These can include video or still cameras, questionnaires to subjects, observations and ratings by researchers (to include recording poor postures, potential accidents, mistakes and so on by the subjects), free comments by the subjects as well as quantitative measures such as the time taken to complete a task or the number of errors made. In the previously mentioned study of door locks for use by older people, the measure of usability was taken to be the time to open the locks in various conditions, including when wearing welding goggles to simulate operation in the dark. An example of a still photograph taken from a video during the testing of harnesses is shown in Figure 7.3. The forward movement of the dummy can be clearly seen.

- It is also wise to have thought about the nature and type of *analysis of the results* that will be carried out prior to the tests. This ensures that the correct data are collected and the implications of the results fully tested statistically.

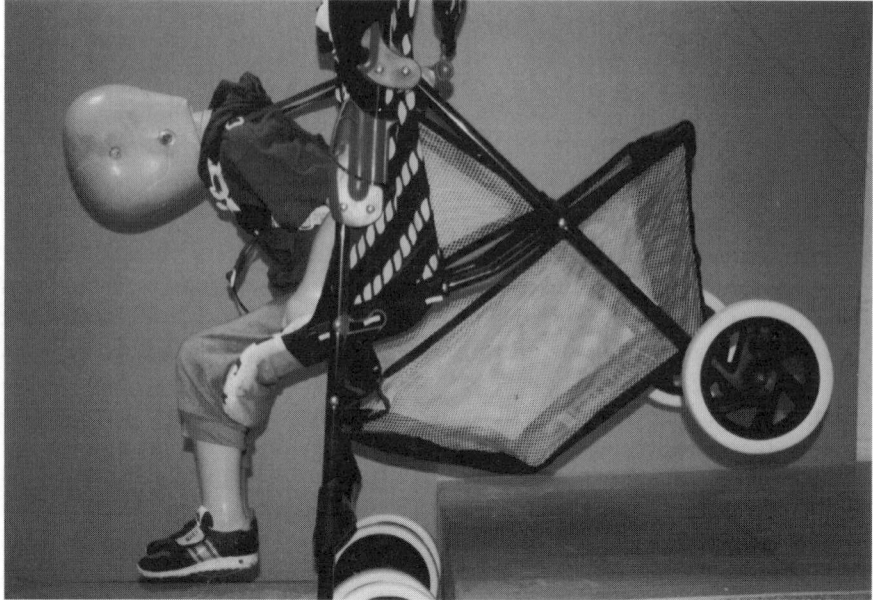

**Figure 7.3**  A dummy 3-year-old restrained in a harness and descending a slope and kerb.

*Interpretation of the Results*

Following the results of any accident analyses and user trials it is normal to draw up a set of performance criteria for the future design or redesign of the product in order to reduce the associated hazards and problems. It is essential to communicate these criteria to those who may benefit, including designers, manufacturers, standards bodies and legislators if they are to have any effect on accident numbers and future product design. Examples of such performance criteria emerged from early work at ICE on lawnmowers, where the following recommendations were made for the redesign of the deadman switch on electric lawnmowers:

- The switch should require two discrete actions by the user to activate the switch initially, not a single gripping action as was present at the time.
- The actions required initially to activate the switch should be convenient to perform – on hover mowers with one hand if they needed to be tilted when starting.
- The switch control should be designed and positioned on the mower handle so that both can be held comfortably throughout the mowing period and at the same time allow comfortable and effective handling of the mower.
- The switch should be designed to discourage artificial circumvention, e.g. by taping in the 'on' position. (Fulton and Feeney, 1983)

In 1994 the recommendations that came out of the study on pushchair harnesses were as follows:

■ There is a need to limit the movement of a child's shoulders and pelvis for reasons of safety, behaviour and parental preference.

■ Manufacturers should supply pushchairs with all necessary means to make them reasonably safe. Parents expect pushchairs to be safe, wish to use a well designed and easy-to-use restraint system and do not want to have to provide additional means of restraint.

■ A dummy was designed for testing the performance of restraints by means of a whole battery of tests. These included a 'wriggle' test to simulate a child trying to shrug its shoulders out of a harness, a tumble test to simulate the pushchair tipping over and various other strength tests on the harness and its adjustment systems.

These practical recommendations can be used by manufacturers to improve the safety and design of their products.

## 7.5  Conclusions

The evaluation of consumer products has an important contribution to make to improving their safety for consumer use. The results that arise from the work feed directly into future product developments as well as enabling injured consumers to obtain redress.

The approaches outlined here are the practical ways in which ICE Ergonomics carries out its research and which have remained viable over the past 25 years. The techniques have proved to be both robust and economical, giving clients the answers they require in a cost effective way.

## Acknowledgement

The author acknowledges the contribution of all staff within the Consumer Products unit at ICE Ergonomics upon whose work much of this chapter has been based.

## Useful Addresses

Child Accident Prevention Trust
4th Floor
Clerks Court
18–20 Farringdon Lane
London
EC1R 3AU

The Health and Safety Executive
HSE Headquarters
St Hughs House
Stanley Precinct
Bootle
Merseyside
L20 3QY

The Royal Society for the Prevention of Accidents
RoSPA, The Safety Centre
Edgbaston Park
353 Bristol Road
Birmingham

## References

DEPARTMENT OF HEALTH (1993) *The Health of the Nation*, Key Area handbook
  *Accidents*. London: HMSO.
DEPARTMENT OF TRADE AND INDUSTRY (1996) *Consumer Safety Unit Report on
  1994 Accident Data and Safety Research*. London: HMSO.
FULTON, E. J. and FEENEY, R. J. (1983) Powered domestic lawnmowers: design for
  safety, *Applied Ergonomics*, **14**(2), 91–5.
HOEFNAGELS, W. A. M., and VAN AKEN, D. (1991) *Safety Criteria for Trapping
  Hazards. Part 1: Inventory and Analysis of Accidents and Existing Standards*,
  Consumer Safety Institute Report 108.
KIRK, N. S. and RIDGWAY, S. (1970) Ergonomics testing of consumer products. 1:
  General considerations, *Applied Ergonomics*, **1**(5), 295–300.
KIRK, N. S. and RIDGWAY, S. (1971) Ergonomics testing of consumer products. 2:
  Techniques, *Applied Ergonomics*, **2**(1), 12–18.
PAGE, M. E., LEE, V. J. A., CLIFT, L. and BIRD, R. J. (1996) Consumers and self-
  assembly products – problems and solutions. Paper presented at the 4th ECOSA
  International Conference on Product Safety Research, Canberra, Australia,
  February 1996.
RENNIE, A. M. (1981) The application of ergonomics to consumer product evaluation,
  *Applied Ergonomics*, **12**(3), 163–8.
SHACKEL, B., CHIDSEY, K. D. and SHIPLEY, P. (1969) 'The assessment of chair
  comfort', *Ergonomics*, **12**(2), 269–306.
WEEGELS, M. F. (1996) 'Accidents involving consumer products', PhD thesis,
  University of Delft.
WILSON, J. R. (1983) 'Pressures and procedures for the design of safer consumer
  products', *Applied Ergonomics*, **14**(2), 109–16.
WILSON, J. R. (1984) 'Standards for product safety design: a framework for their
  production', *Applied Ergonomics* **15**(3), 203–10.
WILSON, J. R. and KIRK, N. S. (1980) 'Ergonomics and product liability' *Applied
  Ergonomics*, **11**(3), 130–6.

# CHAPTER EIGHT

# Enhancing the quality of use: human factors at Philips

PATRICK W. JORDAN[1], BRUCE THOMAS[2] and BRONWEN TAYLOR[2]

[1]*Philips Corporate Design, Groningen, The Netherlands*
[2]*Philips Corporate Design, Eindhoven, The Netherlands*

## 8.1 Organisation

### 8.1.1 Philips

Philips is a diversified global electronics company, with its headquarters at Eindhoven in the Netherlands. The company makes a vast range of products, ranging from consumer electronics to in-car navigation systems, domestic appliances, medical equipment, communications products and lighting systems.

Philips has accepted the need for a user-centred perspective on design, with quality of use as a central tenet. Human factors plays a major part at corporate and strategic levels as well as at product division level in helping to formulate this approach and to implement ergonomic principles in product requirements specification and product design.

### 8.1.2 Philips Corporate Design

Philips Corporate Design (PCD) is the centre of expertise within Philips for the entire range of design activities. The department has 320 members based on 15 different sites in 14 different countries worldwide. Around 150 of PCD's personnel are based in Eindhoven.

PCD is responsible for the design quality of all Philips products. PCD has a model of the design process called High Design. This gives an idealised vision of how the department works. The aim embodied in the High Design process is to integrate basic, traditional design skills with the new design-related expertise needed to respond effectively to the world's increasing complexity and the increasing demands placed on products by the consumer.

PCD works in a context where the organisation of activities is strongly influenced by the product divisions that commission the work. This means that adaptations must be made to the principles of High Design to fit the context of working in a business community. As a result, the approach to design d basically a series of iterative cycles consisting of analysis, creation and evaluation.

A major priority in PCD's work is to tailor products more closely to the needs and characteristics of the users. Human factors plays a central role in helping PCD to achieve these aims.

### 8.1.3   Applied Ergonomics Group

The Applied Ergonomics Group was founded at PCD (then Corporate Industrial Design) in 1986 with the appointment of Ian McClelland as Manager, Applied Ergonomics. The mission of the group was to ensure that the utility of products and that the quality of interaction was appropriate to the requirements of the user and the context in which the product is used. This meant establishing usability as a specified design objective. The challenge was to integrate users' needs, interaction design, and evaluating the usability of solutions into a holistic and iterative design process. The interpretation of usability in this work was explicitly derived from the description given in ISO DIS 9241 *Ergonomics Requirements for Office Work with Visual Display Terminals II: Framework for describing Usability in terms of User-based Measures:*

> the effectiveness, efficiency and satisfaction with which specified users can achieve specified goals in particular environments.

Originally, all human factors specialists at Philips Corporate Design worked within this group which was, thus, a central resource for human factors capacity. However, in 1996 some restructuring occurred so that the majority of human factors specialists left this central group, becoming allocated to design groups dedicated to particular product areas. This restructuring was a direct result of the demand for human factors within Philips – it was felt necessary to have a human factors specialist permanently involved with each design group. A central Applied Ergonomics Group still exists, which provides support for those working in the design groups. This is in terms both of supplying additional effort for development projects, and of developing professional resources for the group, for example information, methods and tools, via a competence development programme.

The work of the group has been described in the past by, for example, McClelland and Brigham (1990) and de Vries *et al.* (1993). There are now 21 human factors specialists within PCD, 6 of whom remain in the central Applied Ergonomics Group.

### 8.1.4 Nature of the work

The emphasis in the work of the human factors specialists at PCD is very much on human factors taking a proactive approach to developing design solutions. Human factors work is firmly embedded in the organisation – 85 per cent of the work is for product divisions. The philosophy of the group emphasises the benefits of integrating user involvement throughout the design process. Understanding the users' experience when they interact with a product is central to developing useful, usable products which will delight the end-user. The application of the skills and techniques to support design decisions promoting quality of use are based as much as possible on measurable indicators. This includes the application of 'usability engineering' principles to the product creation process (see, for example, Whiteside *et al.*, 1988). It also includes considering the less tangible, emotional aspects of product quality (Jordan and Servaes, 1995).

Ensuring effective input to a design means that the human factors specialists work closely with designers, product managers and users at all stages of the product creation process, from requirements specification, through to the evaluation of a finished product. Because the work is carried out within a commercial environment, it is vital not only that ergonomics input ensures a high quality of interaction, but also that it can be provided within challenging time and financial constraints. This means responding speedily and effectively to the demands of commissioners, working to create design solutions that meet the needs of users in a form that can be implemented efficiently within fixed timescales and cost margins.

Because human factors is involved throughout the design process and with a wide range of products, each specialist has to develop a range of skills that can be tailored to a form that is most relevant and efficient given the nature of the product, the characteristics of the user group, the context of use and the stage of the design process which has been reached.

It is recognised that effectiveness can be enhanced by the ongoing development of competence amongst human factors specialists. The Applied Ergonomics Group is involved in a competence development programme to enhance the effectiveness and efficiency of ergonomics input into the design process. This programme gives the opportunity to study the latest developments in ergonomics, as well as the opportunity to conduct original research of their own. Around 10 per cent of the group's time is spent on competence development.

## 8.2   Role of Human Factors in Design

### 8.2.1   Product Creation Process

The user's perspective can be brought in at all stages of the product creation process. Van Vianen *et al.* (1996) describe how different evaluation methods can be used. Issues they address include:

- Purposes of a test
- Specification of participants
- Design of tests
- Measures to be taken
- Questionnaires to be used
- Tasks to be executed
- Presentation of results.

In particular, they show that, at different stages of the product creation process, different evaluation methods can be applied (see Table 8.1). For example, in the early concept creation phase, informal tests can be carried out with a limited number of participants to obtain rapid feedback on the quality of the design and to select a design direction, whereas later in the process a more formal evaluation might be carried out using a larger number of participants drawn from the target population in order to ensure that predefined usability criteria have been met.

**Table 8.1**   Overview of evaluation techniques linked to their purpose and the phase in the product creation process

| PCP phase | Purpose | Techniques |
|---|---|---|
| Know-how | Generation of new concepts | User workshops Focus groups |
| | Inventory of problems | All techniques |
| Concept | Answer specific question | All techniques (related to a specific question) |
| | Compliance with acceptance criteria | Usability test |
| Product range start | Testing final concept | Usability tests |
| Commercial release | Inventory of problems | Inventory |
| | Confirmation of usability criteria | Usability tests |

The Applied Ergonomics Group leads a corporate-sponsored multidepartment, multidisciplinary project aimed at increasing the user and customer focus in product development processes. The project is building a model which defines the key attributes required to ensure adequate user focus. The model is being used to guide process improvements. Included in this project is the development of an assessment tool designed to help product divisions to identify specific improvement targets.

This project has also developed a model user requirements specification which incorporates the usability specification and measurement principles found in ISO DIS 9241 part II. This model has already been written into the process description by one business group as the required method for specifying user requirements in all future product developments.

### 8.2.2 Analysis of User Needs

Bringing the user into the process is not just a question of evaluation. People have needs and desires, which should be reflected in product design. Thomas and de Vries (1995) describe the development process of a business telephone, in which users were not only consulted throughout the process in order to evaluate the design concepts, but also interviewed and observed at their workplaces with a view to understanding their requirements. A more thorough involvement of users in the creation of concepts is described by Oosterholt *et al.* (1995).

The private camera conversation is a good example of a tool developed in-house to investigate the issues of concern to users (de Vries *et al.*, 1996). This method involves participants entering a private booth and talking to a video camera about a predefined subject identified by the investigator. Participants might talk about the way in which they use a particular product, how easy or difficult a product is to use, or how a product fits into their way of life. Because participants are only given a very general brief regarding what they should talk about, the issues covered are likely to be almost totally driven by the participant. People often find the method fun to take part in and there is the additional benefit of having no investigator/participant interaction effect. This can lead to participants being more open. However, it is difficult to put any controls on this method and there is no guarantee that participants will cover the issues that are of interest to the investigator. A variant on the method is to have two people speaking to the camera at the same time. This can have the advantage that participants prompt each other by picking up on points that the other has made.

A further technique has been developed to investigate the relative merits of particular product features: the valuation method (Jordan and Thomas, 1994). This method involves asking participants how much extra they would pay for a product if the product were to contain particular features. The method can be particularly useful during requirements capture as a way of comparing the potential benefits of different features. This can help in making a trade-off in situations where there are limits to the functionality that can be included. Asking users about how much they would pay for something is a way of anchoring responses in a fairly concrete context. People are used to making monetary value judgements in their daily lives and may thus find this easier than making somewhat abstract judgements, such as might be required when having to choose between whether a feature is, say, important or very important on a Likert scale. The method also provides quantitative comparative data which is likely to be easier to analyse than responses to open-ended questions about how important something is, which will require more interpretation in analysis. A limitation of the method is that it would not be wise to interpret the responses given as being truly representative of what people would pay. Obtaining realistic estimates of this facet is notoriously difficult. The figures gained from the Private Camera Conversation are solely for the purpose of comparison between features.

### 8.2.3    Interaction Design

A main approach to interaction design taken by the human factors specialists at PCD is to focus on the tasks that the user will carry out. This is particularly the case for products to be used in the professional area, where performing tasks is the fundamental aspect of product use. In the consumer area, performing tasks may be a secondary aspect of product use, where, for example, entertainment and other less tangible factors play an important role. Nevertheless, in the design of interactions, focusing on the goal of activities leads to the development of solutions which users, whether professional or consumer, should find satisfying. The process of developing an interaction design is shown in Figure 8.1.

The first step in developing an interaction is, therefore, to identify the main user tasks, i.e. what the user will want to do with the product. This step should be technology-free to enable the interaction designer to consider novel solutions. Thus, the designer is not constrained by the way that things have been done in the past, and is not forced to repeat the mistakes made in the past. For example, to develop a simpler method of programming a video recorder, the focus is on the user's desire to 'record *Star Trek* tonight', rather than the operational demands of setting a start time, an end time and a TV channel. Further, this technology-free consideration is then valid not only for the product of immediate concern but also for several models. It can also lead to the development of totally new product concepts.

The technology available must be considered at the second step – where the operations are identified – which the user must perform to achieve the desired goals. Here the limitations and advantages of the technology available must be considered from the point of view of obtaining the optimal interaction

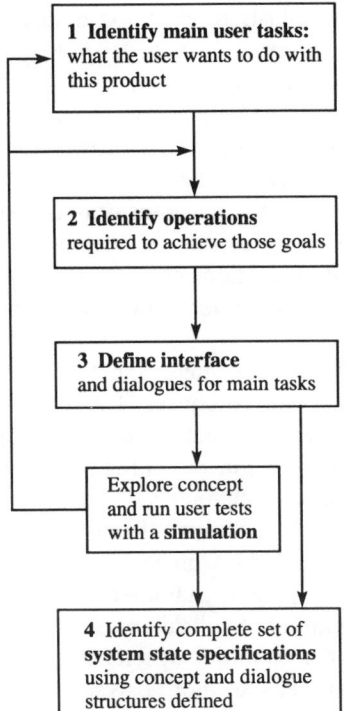

**Figure 8.1**    The interaction design process.

structures. Where basic technology remains the same, this may require only small changes from previous models.

The third phase of interaction design is to define the interface and to generate dialogues for the main tasks. This covers all the main dialogues but still takes a user-centred view by maintaining a focus on the tasks that the user performs rather than the functions that the product offers. At this stage the feedback provided and the cues given to the user are described using diagrams, screen layouts, etc.

At this point in the design process it becomes possible to conduct an initial evaluation of the concepts. It is preferable to simulate the interactions, even at a fairly primitive level, in order to obtain confirmation of their suitability. At this stage an iterative loop opens, which enables the designer to go back to the definition of operations, enhance and improve them, redefine the interface and so on. The tools available for evaluation are discussed in section 8.2.4.

After a number of iterations, depending on time and budget available, and on the acceptability of the final concept, the interaction specification can be written. This identifies the complete set of user interface states, using the concept and dialogue structures defined. The specification will usually consist of text plus diagrams plus illustrations of the display format and information content. There may be a variety of formats depending on who will be using this document. It may contain flow diagrams, or a specific format used by the software engineers such as state transition diagrams. In some cases, a full simulation of the interface is used as a specification tool.

A task-based tool, developed in-house, called a Task Action Description (TAD) has been developed for the purposes of documenting interaction design. This tool has been used successfully in the development of a number of products, including a communications control interface for emergency services and other privately operated communications systems (Thomas and McClelland, 1996).

The TAD describes the system states and the transitions between them in terms of user actions and system responses. The form of the TAD was designed to be transferred directly on to state transition diagrams, which was the preferred tool of the software engineers in the team. Thus, the TAD is tailored to *its* users, namely the software engineers.

The concept of the TAD is derived from the ideas of hierarchical task analysis developed originally by Annett and colleagues (see Annett and Duncan, 1967; Annett *et al.*, 1971; Stammers, 1996), in which the operators' tasks are divided into a number of subsidiary objectives and actions required in order to achieve them. An essential difference between the TAD and task analysis, however, is that tasks are not analysed but created. The key idea is that the tasks are structured to facilitate the achievement of particular operator goals in a manner which is consistent with the operators' working practices and expectations.

### 8.2.4    Evaluation

Evaluation has an important part to play in the product creation process. Through evaluation, information is generated which is used to steer later iterations of the development as well as to determine whether the final design will actually meet user needs.

Van Vianen *et al.* (1996) identify three basic questions which must be addressed before conducting any evaluation:

- Are we trying to find out what people want with regard to user interfaces?
- Are we trying to find out what people accept with user interfaces?
- Are we trying to find out what people are able to use?

These questions identify the purpose of a test and will influence the choice of methods and techniques to be used.

Much of the group's usability evaluation work is conducted in a usability laboratory based on-site in Eindhoven. The design of the laboratory was specified on the basis of a requirements capture involving all the group's ergonomists and a sample of the group's commissioners. This has led to a facility which supports effective and efficient evaluations and which can be adapted to commissioners' needs. Care has been taken to design the laboratory such that the environment is pleasant and relaxing for test participants. The laboratory is used not only to evaluate usability but also to create appropriate design solutions using techniques such as focus groups, user workshops, co-discovery sessions, interviews, etc. The laboratory and the work that it supports have been described by de Vries *et al.* (1994).

A range of methods are employed to evaluate designs and products. The selection of method depends on the requirements and on the stage of the creation process. Furthermore, the methods employed not only address usability in the conventional sense but also take into account cultural issues, range of contexts, variations in user populations, pleasure in product use, etc. This is a wide variety of demands, which is met by a variety of methods in regular use, including both qualitative and quantitative as well as formal and informal methods. In addition to standard methods employed by human factors specialists, a number of tools have been developed in-house to address the particular needs of working to the tight constraints imposed in an industrial context.

In the context of a commercial environment, ergonomists in industry often have to trade scientific rigour against achieving timely and usable results. The ergonomist has to produce a definite recommendation to enable designers and developers to continue their work, even when comparative tests have proved inconclusive. Commercially, it is also important to take account of the negative responses to products when they are intended for a wide range of consumers. Ergonomists at PCD have been using a simple decision aid which utilises both the positive and negative data gathered during a test – the Residual Acceptability Index (RAI – see Taylor *et al.*, 1995).

The RAI is intended for use with data gathered on rating scales or from ranking exercises. It is calculated by adding the frequencies of all positive scores and subtracting the sum of the frequencies of negative scores. The remaining figure represents the residual acceptability (which can be negative). Because the RAI is calculated by subtracting the negative scores from the positive, it is a measure of the safety of each alternative.

The short and intense product development cycles in which PCD works significantly reduce the opportunity to carry out usability tests. A solution that has been applied with success is to carry out informal 'quick and dirty' usability tests (Thomas, 1996). These tests can be described as stripped-down versions of formal usability tests, which are based on the measures of usability defined in ISO DIS 9241 part II. The principal compromises made concern the match of the participants in the test to the target population and the precision of the measures used. This does raise questions about the validity of the test results, but PCD has found that such testing can reveal potential problems and provide measures of success which are of benefit in further product iterations.

## 8.3  Relationships

Jordan *et al.* (1996) point out that a key to becoming integrated in the design process is the ability to work in multidisciplinary teams. A role of the human factors specialist in such teams is to provide reflection on design: usability evaluation can be seen as an opportunity for doing this. The usability practitioner can also provide information about users' requirements and ideas for design directions. In order for human factors issues to be recognised and designed into products, the schism of human factors specialists versus designers must disappear. Designers are not beings from another planet. Working together with designers to develop usable solutions should ultimately be more effective than taking a role solely as an evaluator, as this might make the usability practitioner appear rather aloof.

The need to work with other professionals is not limited to designers alone. Others with a responsibility for customer satisfaction are also involved. There is a growing need for co-operation with marketing, sales and product management, and for people in these disciplines to recognise the role that a human factors specialist can play. One simple device to promote co-operation is to involve product managers as participants in usability evaluations – this can operate as an effective propaganda tool, giving them a feel for what it is like to be on the 'receiving end' of the design decisions taken.

Thomas and McClelland (1996) highlight three issues with regard to co-operation between disciplines:

- the importance of direct, first-hand exposure to the daily working environment of the operators by the team as a whole;
- the problem of how to articulate operator requirements in such a form that they are taken into account in the design development;
- the progressive evolution of organisations and changes in work practices.

The success of Philips' design work is attributable to the close working relationships developed between the members of development teams, which can include product management, user interface designers, software and mechanical engineers, marketing experts and product designers. The establishment of close working relationships in such teams enables Philips to create a match between conceptual models, dialogue structures, underlying software and graphic design. Ultimately, we believe this co-operation, along with an understanding of the commercial and industrial processes involved, is a fundamental part of ensuring that the products that Philips brings to the market not only meet but exceed user expectations.

# References

ANNETT, J. and DUNCAN, K. D. (1967) Task analysis and training design, *Occupational Psychology*, **41**, 211–21.

ANNETT, J., DUNCAN, K. D., STAMMERS, R. B. and GREY, M. J. (1971) *Task Analysis*. London: HMSO.

DE VRIES, G., THOMAS, D. B. and MCCLELLAND, I. L. (1993) De Applied Ergonomics Group van Philips Corporate Design, *Tijdschrift voor Ergonomie*, **18**(5), 22–5 (in Dutch).

DE VRIES, G., VAN GELDEREN, T. and BRIGHAM, F. R. (1994) Usability laboratories at Philips: supporting research, development and design for consumer and professional products, *Behaviour and Information Technology*, **13**(1), 119–27.

DE VRIES, G., HARTEVELT, M. and OOSTERHOLT, R. (1996) Private camera conversation: a new method to elicit user information? *Behaviour and Information Technology*, **14**(6), 358–60.

JORDAN, P. W. and SERVAES, M. (1995) Pleasure in product use: beyond usability. In: Robertson, S. (ed.), *Contemporary Ergonomics 1995*. London: Taylor & Francis.

JORDAN, P. W. and THOMAS, D. B. (1994) But how much would you pay for it? An informal technique for setting priorities in requirements capture. In: Robertson, S. A. (ed.), *Contemporary Ergonomics 1995*. London: Taylor & Francis, pp. 145–8.

JORDAN, P. W., THOMAS, D. B. and MCCLELLAND, I. L. (1996) Issues for usability evaluation in industry: seminar discussions. In: Jordan, P. W., Thomas, B., Weerdmeester, B. and McClelland, I. L. (eds), *Usability Evaluation in Industry*. London: Taylor & Francis, pp. 237–43.

MCCLELLAND, I. L. and BRIGHAM, F. R. (1990) Marketing ergonomics: how should ergonomics be packaged? *Ergonomics* **33**(4), 391–8.

OOSTERHOLT, R., KUSANO, M. and DE VRIES, G. (1995) Interaction design and human factors support in the development of a personal communicator for children, *CHI 1996 Conference Proceedings*. Reading, Mass.: Addison-Wesley, pp. 450–7.

STAMMERS, R. B. (1996) Hierarchical task analysis: an overview. In: Jordan, P. W., Thomas, B., Weerdmeester, B. and McClelland, I. L. (eds), *Usability Evaluation in Industry*. London: Taylor & Francis, pp. 207–13.

TAYLOR, B., THOMAS, B. and NEERVOORT, P. (1995) The Residual Acceptability Index (RAI): when you simply must make a decision. Paper presented at the Conference 'Ergonomics in Consumer Product Design', Southampton, 10 November 1995.

THOMAS, B. (1996) Quick and dirty usability tests. In: Jordan, P. W., Thomas, B., Weerdmeester, B. and McClelland, I. L. (eds), *Usability Evaluation in Industry*. London: Taylor & Francis, pp. 107–14.

THOMAS, B. and MCCLELLAND, I. L. (1996) The development of a touch screen based communications terminal, *International Journal of Industrial Ergonomics*, **18**, 1–13.

THOMAS, B. and DE VRIES, G. (1995) A user centred approach to the development of business telephones. In: *Proceedings of the 15th International Symposium on Human Factors in Telecommunications, Telecom Australia, Melbourne*.

VAN VIANEN, E. P. G., THOMAS, B. and VAN NIEUWKASTEELE, M. (1996) A combined effort in the standardisation of user interface research. In: Jordan, P. W., Thomas, B., Weerdmeester, B. and McClelland, I. L. (eds), *Usability Evaluation in Industry*. London: Taylor & Francis, pp. 7–17.

WHITESIDE, J., BENNET, J. and HOLZBLATT, K. (1988) Usability engineering: our experience and evolution. In: Helader, M. (ed.), *Handbook of Human–Computer Engineering*. Amsterdam: Elsevier, pp. 791–817.

# CHAPTER NINE

# Consumer product evaluation: which method is best? A guide to human factors at Consumers' Association

LINDSEY M. BUTTERS

*Consumers' Association Research and Testing Centre, Milton Keynes*

## 9.1 Take Twenty Kettles: Which is the Easiest to Use?

What is the best method for gathering ease-of-use data on large numbers of products? To what extent does the product itself influence your choice? How do you ensure validity, reliability and auditability of results within cost constraints and tight deadlines? How do you present a wealth of subjective data to a client who is used to dealing only with figures?

Based on the experiences of the human factors department at Consumers' Association Research and Testing Centre, this chapter suggests answers to these questions and offers practical guidance to those involved in consumer product evaluation. While it focuses on the production of post-market comparative data for use as a buying guide, much of the advice can be applied equally well by those involved in the iterative design of new products.

Consumers' Association (CA) is committed to empowering people to make informed consumer decisions and to achieving measurable improvements in goods and services. At the CA Research and Testing Centre, a vast range of products is tested, from cars to garden spades, for reporting in *Which?* magazine and CA's other consumer publications. The primary aim is to test the products in a way which accurately reflects consumer usage behaviour.

Technical performance testing, including safety checking and reliability research, is complemented by assessments of the ease of use or *convenience* of a product. A convenience assessment encompasses all aspects of living with one's choice, including familiarisation, the appropriateness and ease of use of the functionality, comfort, anthropometric fit, adjustability and ease of maintenance.

The chapter comprises:

- An overview of CA's approaches to gathering convenience data.
- A product guide to choosing a method.
- Some do's and don'ts for maximising data quality, including getting the most out of a user trial and the design and administration of convenience checklists.
- How to update and validate methods.
- A case study on the evaluation of consumer products for people with special needs.

## 9.2 Evaluation Methods: An Overview

CA's choice of methodology is governed by a number of factors including the following:

- The characteristics of the product (especially safety, complexity, size and monetary value).
- The frequency with which that product is tested.
- Time and cost constraints.

Ideally, the first information about the convenience of a new product type should be gleaned from a user trial. This provides the opportunity to gather both subjective and objective data about the experiences of representative consumers in using the product. Our trials may involve either lay or trained panels of users, both in the user's home and at the CA testing centre and gardening sites.

Home user trials are conducted when the domestic realism of the testing is considered to be an important determining factor. This approach allows the users plenty of time (usually one week per sample) to familiarise themselves with the product in their own surroundings and to test it thoroughly. The success of a home trial relies upon users doing as they are asked and giving each of the samples a fair trial. Laboratory or field trials facilitate observation of the users and careful control of the testing environment.

For some products, once CA has built up a body of knowledge from user trials about consumer behaviour and product usage, a convenience checklist is devised for use on future projects. Incorporating relevant standards, our checklists include many of the questions that may be asked of a user in a self-completion questionnaire. Checklists are completed by a team of trained laboratory staff, who each rate the product separately before discussing their findings to reach a consensus. Guidance on good ergonomics practice (e.g. on recommended control dimensions) is included on the checklists. Such checklists are particularly appropriate for products which are tested frequently or are too complex for all the functionality to be tested in a single user trial.

An expert appraisal by CA's ergonomist may be used, particularly if an assessment of compliance with standards is required.

## 9.3    Choosing a Method: A Product Guide

### 9.3.1    Domestic Appliances

Small domestic appliances which require no installation, such as kettles and vacuum cleaners, are ideally suited to a home user trial. Such products can be easily delivered to and collected from the user's home. As they are used regularly in the home, we do not have to ask the users to do much in addition to what they would be doing anyway.

These are also products which are likely to be strongly affected by the realism of the testing environment. For example, a vacuum cleaner may be very efficient at sucking up standard dust in a laboratory test but be so cumbersome that users have difficulty moving it around furniture or carrying it upstairs.

Large 'white goods' (such as washing machines, tumble driers, dishwashers and refrigeration products) are not so suited to home trials. They are difficult to install and expensive to buy (a home user trial requires several samples of each product or 'code' if it is to be completed within reasonable time). Therefore, on the basis of laboratory user trials, convenience checklists have been developed for these products.

### 9.3.2    Cars

Car testing at CA is an example of the use of a trained panel. The in-house panel consists of staff who are ordinary motorists and consumers, all shapes and sizes with assorted patterns of use and driving styles, who are experienced in assessing the cars for a huge number of factors. These include ergonomics issues such as dashboard layout, driving position, visibility, etc. The assessments are carried out as if the car were their own (e.g. for travelling to and from work, trips to the supermarket, weekend outings, etc.). Having driven the car for a week, on all road types, they fill out a detailed questionnaire covering all aspects of the vehicle. They provide information on aspects such as inadequate headroom for a tall person, or cars that can be annoying to drive in wet weather because the windscreen wiper does not clear enough of the screen. The results are analysed statistically and combined with technical measurements and track tests to give overall ratings. A detailed profile of the characteristics of the drivers enables us to interpret their comments in the light of anthropometric considerations.

### 9.3.3   Audiovisual and Telecommunications Products

For many of the products in this area, the learning curve is considered to be too steep to enable the products to be evaluated effectively by means of a user trial. In these cases, our checklists are drawn up by ergonomists in the light of existing guidelines and in consultation with the staff responsible for performance testing. Supplementary user trials can be employed to investigate consumer reactions to particular aspects or features.

The ease of use of some products is very 'feature-dependent', i.e. inclusion of a particular feature will automatically make for greater convenience (outweighing the *implementation* of that feature). Certain types of camera fall into this category. A very objective, quantitative checklist is possible, with a series of tick-boxes and scores for particular features.

### 9.3.4   Software

From within CA's user panel, subgroups can be identified who are known to have the right aptitude, interest and equipment to carry out software evaluations. Users are given the software to test in their own homes. For example, in a recent study of home finance packages, users were recruited who were already accustomed to using their computer to manage their own accounts. They tried out the packages (with a mixture of prescribed tasks and unstructured use) for two months.

### 9.3.5   Gardening and DIY tools

Safety is often the most important factor here. When compost shredders first appeared on the market, CA did not consider them safe enough to be assessed by user trial and instead opted for a checklist approach. As standards have improved, CA has felt able to run an observed user trial but has not yet sent out shredders for testing at home.

Observed trials are possible even if carried out at the 'wrong' time of year. On one occasion, scheduling demands meant that CA was assessing weeding tools in the winter, ready for reporting in the spring. There are fewer weeds to dig up in the winter and it is not the right time to ask people to be out in their gardens. Therefore, weeds were grown under a polythene tunnel at a garden test site and keen gardeners were able to test the tools under cover.

A highly objective checklist is appropriate for many gardening products. For example, CA's assessment on mowers uses predefined rating categories according to the weight of the mower, where the starter control is positioned and so on. As always, these are based on the results of user trials.

An observed user trial is used in preference to a home trial if normal use of the product is infrequent. For example, if people were given DIY eye protectors or workbenches to try out at home they would be unlikely to have sufficient suitable tasks over the course of a week to give them a fair test.

## 9.4  Maximising Data Quality: Some Do's and Don'ts

### 9.4.1  Getting the Most out of a User Trial

*Defining the User Profile*

It is important that the user profile (including age range, male/female ratio and knowledge of the product on test) is clearly defined and understood from the outset. It is difficult to get truly representative users, especially if they are self-selecting: but the more one knows about the characteristics of the user group, the better one can allow for any biases which may be present, interpreting the findings accordingly.

A panel of nearly two thousand users in just over one thousand households, all of whom are subscribers to one or more of CA's magazines and all living within twenty miles of the Research and Testing Centre, has volunteered to participate in the trials. For all panel members, CA has a wealth of background data concerning their household appliances, their skills and interests; these are held on a database to assist in identifying appropriate user samples.

Careful recruitment is needed to select users with a suitable level of knowledge and expertise. For example, the more specialist woodworking tools can only be properly assessed by people with appropriate skills. On the other hand, products such as software aimed at the home market need to be usable by a wide range of people, not just those who are skilled in the use of high technology. However, it would be inappropriate to recruit completely naive users with no experience of using computers.

Occasionally, a specialist group of users is needed (for example, a cycling club to try out racing cycles or very young babies to try out disposable nappies). On these occasions, CA recruits from outside its established panels.

*Importance of Briefing Sessions*

Irrespective of the testing location, briefing users as a group rather than individually instils a sense of teamwork. Users are more likely to complete the trial conscientiously if they understand the bigger picture and that their contribution affects the overall quality of the results.

## Do Not Overload the Users

For a home trial, one sample per week over a four or five week period is the maximum advised if users are to retain their interest. An observed trial in the laboratory usually lasts between two and three hours, with the number of samples tested depending upon the nature of the tasks.

## Maximising Realism

Observed trials inevitably create an artificial situation, but this can be alleviated by imitating the most important characteristics of the domestic environment. For example, when users test wheelbarrows they are provided with a mocked-up course containing the sorts of obstacle which they might expect to encounter in their own gardens. Users testing eye protectors perform a series of controlled tasks, including painting and sanding walls, to simulate real-life use.

## Data Capture and Design of Questionnaires

Usually, data from user trials is captured through the use of self-completion questionnaires (one for each product tested) and the results are analysed statistically. Such questionnaires are unobtrusive and relatively easy to administer. Occasionally, only anecdotal information is required. In all cases, testing is conducted according to an experimental design to ensure a fully randomised, balanced order of testing. Where the trial is to be analysed statistically, numerical rating scales with a midpoint are used; a five-point scale is usually adopted. Space for comments is allowed and users are encouraged to list any particularly good or bad features which contributed towards their overall assessment. The rating scales give easily analysed but non-specific data, while the written comments are more specific but harder to analyse. In an observed trial, the information provided by users can be supplemented by recording features of the session such as the time to complete tasks, causes of user difficulty, errors made, and the user's approach to using the product. An important question to ask is whether users would choose the product in preference to the one that they currently own. Data showing voluntary usage is really the ultimate subjective satisfaction rating.

Videotaping the trial is worth considering for future reference and to allow a fuller analysis of the interaction than is available by direct observation (but recognise that the recording may interfere with users' performance and tape analysis is very time consuming).

## How Many Users Do You Need?

The answer to this depends upon the type and quality of data required. In much of CA's testing, where it is necessary to be sure that any conclusions drawn are statistically significant and have taken into account individual differences between people, quite large numbers of users are needed. Typically, twelve sets of results are generated for each product or 'code' tested. Given that a project can include anywhere between 10 and 40 codes, dozens of users are often needed.

Anecdotal information, on the other hand, can be collected using quite a small group. According to Virzi (1992), 80 per cent of usability problems are detected with four or five subjects, additional subjects are less and less likely to reveal new information, and the more severe problems are likely to have been detected in the first few subjects. Separate studies by Lewis (1994) clearly support the second claim, partially support the first, but fail to support the third. Gould and Lewis (1985) suggest that user diversity is overestimated and that the same problem, even a completely unanticipated one, often crops up for user after user.

It is CA's policy to consider the needs of all users, small and tall, young and old, able-bodied and those with special needs, but it is not always possible to accommodate all of this within one trial. Supplementary trials with small numbers of users can be an effective way of providing anecdotal information to help people with particular requirements. For example, CA included a 'left-handers' session when testing corkscrews. A recent trial on secateurs was supplemented by sessions involving users with poor grip. The 'best buys' from the main trial were compared against brands which claimed to be especially suitable for those with arthritis. (An approach to the systematic collection of more detailed convenience data for people with special needs is outlined in the case study at the end of the chapter.)

*Discussion Groups*

Users may be recalled to participate in discussion groups as a debriefing exercise at the end of a user trial. A great deal of valuable information can be collected in a short period of time which can be used to supplement and to help interpret the statistical findings. The technique is particularly good for obtaining ideas and recommendations for improvement. However, it demands skilled administration to ensure that all opinions are represented and not simply a consensus based on 'middle ground'. For this reason it often works best when the discussion group is composed of a homogeneous sample of users. Some users may be inhibited in a group situation and be unwilling to admit negative aspects of their interactions. On the other hand, sessions may be hijacked by a particularly vocal participant.

Where a mini-trial is envisaged – to investigate a subset of complex functionality – discussion groups can be a useful means of helping to decide which aspects of the product should be concentrated on. In this case, the group discussion would be held before the user trial or after a pilot trial has taken place.

### 9.4.2  Design and Administration of Convenience Checklists

Checklists facilitate the speedy production of numerical data through the use of tick-boxes and rating scales. They must be carefully designed and tightly controlled if reliability is to be maximised and maintained, and should include representative tasks on which to base the assessment.

*Getting the Balance Right*

Checklists must be detailed enough to capture all relevant data and to give sufficient guidance on how the checklist is to be administered, without being too long-winded and time-consuming to complete. They must be prescriptive and precise enough to help the assessors award a fair and consistent rating, and for all testers to arrive at similar, if not identical, answers (otherwise the final rating may be attributable more to experimenter whim than anything else).

*Defining the Rating Scales: What Do the Poles Represent?*

If assessors are asked to rate the products on, say, a five-point scale, they need an explicit definition of the scope of that scale. For example, is a score of 5 to be awarded to the best of what is available or is it an 'ideal' which has not yet been designed? By adopting the former option, you risk having to shift your scale as products improve (to allow results from previous projects to be compared fairly). With the latter, you 'squeeze' the scale and achieve less discrimination between products because the minimum and maximum scores are rarely awarded.

*Using Objective Criteria*

It is often easier for testers to make comparative assessments by collecting *objective* data (e.g. measuring the size of a control, the force required to turn it, etc.). However, interpretation of such measurements is often difficult because many of the available data and recommendations in ergonomics guidelines are incomplete, out-of-date and do not reflect the developments in modern domestic products. For this reason, the emphasis in the majority of CA's checklists is on subjective assessment. The exceptions are those products, already described, for which ease of use is very feature-dependent.

*Training Issues*

There is a considerable training load in equipping laboratory staff with the skills needed to conduct ease-of-use assessments. Moreover, it is debatable whether, even with training, a male assessor, for example, can gauge whether a particular control is easily adjusted by a female, a child or the elderly. For this reason, CA's policy is for the checklists to be completed by both male and female staff, and for a left-handed assessment to be included.

## Limitations

It is important to be aware that expert laboratory staff, who are familiar with the products having tested them for technical performance, do not relate to those products in the same way as does the ordinary consumer. Their expert knowledge puts them in an excellent position to give detailed advice on what is good or poor about a product, but there is a downside to this knowledge in that they become almost 'too expert': it is very difficult to think about whether or not someone else will have trouble if you never encounter any yourself.

Checklists have speed of completion as their main advantage, but their apparent simplicity in comparison with user trials can be misleading. They are very good for obtaining information on the current product and its strengths and weaknesses, but do not necessarily capture those features that are missing.

### 9.4.3   Supplementing the Data Using Other Methods

## Expert Appraisals

An expert appraisal by CA's ergonomist may be used to predict how a product is likely to suit a range of users over time, to assess compliance with a standard and to identify design flaws which may lead to problems. The usefulness of the method hinges on the expert's experience with the product and knowledge of the user's likely behaviour. However, as with laboratory staff, this experience can be a double-edged sword: it can be very difficult to set aside one's expertise and assume the role of a user. For this reason, it is a method which is rarely used in isolation but is more commonly used to complement other methods. For example, the usability element of a report on electronic route-finders (*Which?*, 1996) comprised a user trial plus an evaluation of the user interface by the ergonomist.

Comparison with a standard is an obvious way of making comparative assessments. However, standards generally apply to only a limited aspect of a product's usability (e.g. keypad feel and dimensions, legibility of displayed information, etc.). There are also a large number of standards and design guidelines in existence, many offering conflicting recommendations.

## Convenience Diaries

The learning curve is an aspect of convenience testing with widespread implications. It can be difficult to judge whether the aspects that seem inconvenient on first acquaintance with a sample would remain inconvenient on prolonged use. Conversely, inconveniences which at first sight seem minor can become more irritating with time. This is one of the advantages of home user trials over on-site trials because the user can spend much more time with the product. Convenience diaries, kept during performance testing as a back-up source of information, can also be useful in assessing those aspects which are likely to be a source of long-term annoyance.

## 9.5  Presenting the Data

Within the constraints of limited magazine space, CA aims to give people the information with which to make an informed decision. For example, reporting on a car as having controls that are a 'stretch' for smaller users encourages people to check this aspect for themselves when buying. CA advises its readers on the types of controls, handles, etc. to look out for and highlights those products that are particularly good or bad in each area of design.

Many of the data are published as five-point rating scales, which require careful interpretation to enable the reader to draw out the differences between products and to decide what the analysis means for them. We identify products that are best for particular groups, such as left-handed users, and those features which make a product particularly useful for the elderly, the visually impaired or those concerned about child safety. CA also aims to distinguish between genuine advances in ease of use and cosmetic sales techniques, a factor which is becoming increasingly pertinent as more manufacturers are starting to recognise the potential of ergonomics for market advantage.

## 9.6  Updating and Validating Methods

CA's evaluation methods vary not only from product to product but also between different assessments of the same product. We take into account the convenience history and look at which methods have been used in the past, at where the product is in its lifecycle and at our frequency of testing. Only by subjecting the methods to constant review and refinement is it possible to ensure the highest levels of reliability, validity and auditability.

### 9.6.1  Periodic User Trials to Validate and Update Checklists

Given CA's strong commitment to the philosophy of testing products as the consumer would use them, it is important to submit the checklists to regular reviews. This ensures that all relevant changes in legislation and standards are included and that the checklist reflects the state-of-the-art of the product concerned. Design tweaks and seemingly minor model changes may have a significant cumulative effect: a case of the whole being much greater than the sum of its parts.

The principal method of investigating changing patterns of consumer behaviour or the effects of changes in product design is by means of supplementary user trials. For example, lawnmowers are a product traditionally assessed by checklist, but a user trial is periodically conducted in parallel. The degree of correlation between the two sets of results is analysed and any necessary adjustments to the checklist are made.

### 9.6.2   Comparison of Methods: Do the Results Tell the Same Story?

A preferred method can be honed and validated by carrying out a battery of different methods on the same product. Where the results converge, this is a good indicator of their validity. Where they diverge, the discrepancies must be examined to identify possible areas of improvement.

Different methods can sometimes produce differing results without this necessarily meaning that one set of results is more correct than another. Differences are most likely in a complex product which has a steep learning curve, such as video recorders. There is likely to be a conflict between ease of *first-time* use (as judged on a one-off trial by users) and convenience on a *day-to-day* basis (as assessed by experienced testers). It is better to employ user trials to investigate one aspect of complex products: aiming for a detailed assessment of one aspect is much more productive than a more cursory look at the whole product. By concentrating on a subset of the most commonly used functionality within a user trial, it is possible to ensure that CA is getting the basics right and continuing to report on the issues which are most important to consumers.

Even a simple product, although more likely to produce convergent results, can benefit from such an exercise. For example, a convenience checklist on kettles, when compared with results from a home user trial, will reveal the extent to which domestic realism (such as the shape of the sink, design of the taps, etc.) affects the results for ease of filling. We can then make a judgement about whether a checklist is appropriate for this product.

### 9.6.3   Consumer Surveys

A further element in the iterative improvement of our methodology involves employing the resources of CA's Survey Centre to improve our understanding of people's use of consumer products and the problems they encounter.

For example, questions on the ease-of-use of vacuum cleaners were included in the 1995 *Which?* members' annual survey. This is a survey which is sent to 50 000 members, randomly selected to be representative of our membership.

Typically, it achieves a response rate of between 40 and 50 per cent. It is used to generate subsamples according to particular topics for further surveys during the course of the following year.

In the case of vacuum cleaners, respondents were asked to choose five features from a prescribed list of eighteen items which were the most important in their choice of vacuum cleaner. The list included performance, aesthetic and convenience factors. In a second question, respondents were presented with the same list and were asked to choose those five features which they felt were most in need of improvement, from their experience of using the vacuum cleaner. By comparing the two sets of responses, we could look at people's attitudes to convenience issues before and after purchase, influences upon the buying decision and the main problems encountered since purchase.

This information feeds into our principles for the collecting and reporting of usability data within CA and acts as a check that we are reporting on the most important convenience issues.

## 9.7  Case Study: Evaluating Consumer Products for Those with Special Needs

Consumer products have, by their very nature, to cater for the broadest of user groups. Over six million adults in Britain have some level of disability. As we get older, many of us start to experience difficulties with our sight, hearing, mobility and the use of our hands.

*Which?* articles have traditionally included specific information for less-able consumers. Much of CA's work in this area is conducted in collaboration with the Research Institute for Consumer Affairs (RICA), a sister charity to Consumers' Association specialising in research on behalf of elderly and disabled consumers. This collaboration has led to the development of a fully integrated assessment method whereby each product is assessed for how well it meets the needs of any user, whatever their ability.

It is an approach based on the premise that those products which are the easiest to use for consumers with special needs are also likely to be the most convenient for the able-bodied. By adopting more critical assessment criteria, it is possible to produce data which discriminates between products more effectively for *all* consumers.

Developing the checklists which form the basis of the assessments has been the culmination of many years' work. As a starting point, product-specific user trials were commissioned, involving users with a wide range of disabilities (including blind and partially sighted people, wheelchair users and those with other mobility problems, and people with co-ordination difficulties, dexterity problems and limitations of reach and strength). The trials were tightly controlled and run by disability experts to focus the trials on the most important issues and to minimise the element of clinical risk: the safety of users is always paramount and never more so than when those users are elderly or disabled.

The knowledge gained formed the basis for special needs convenience checklists administered by laboratory staff who have undergone specific training to assess products for those with special needs. Initially, these were used alongside CA's 'able-bodied' checklists: the two have now been combined to gather all the data in one assessment.

Concentration is currently focused on those domestic appliances which are most important to independent living, such as washing machines and vacuum cleaners. A question for the future is whether the approach can be extended to other product groups. Given appropriate training and tightly structured checklists, it is reasonably easy for able-bodied testers to assess, say, a washing machine control for use by those with poor grip. A large dial with a knurled edge is going to be easier to turn than one which is small and slippery: the assessment can be made mainly by observation. However, in a project on garden spades, for example, it is likely to be much more difficult for an able-bodied tester to judge how easy it is to dig when suffering from a bad back. It may be that a checklist is possible for some products and for some categories of disability, but not all. Those products for which a checklist is deemed inappropriate would require a special needs user trial on each project.

Through a more stringent aim of testing for *all* consumers, including those with special needs, CA and RICA are seeking to report upon and influence the willingness of manufacturers to embrace 'design for all'.

# References

GOULD, J. D. and LEWIS, C. (1985) Designing for usability: key principles and what designers think, *Communications of the ACM*, **2**(3), 300–11.

LEWIS, J. R. (1994) Sample sizes for usability studies: additional considerations, *Human Factors*, **36**(2), 368–78.

VIRZI, R. A. 1992, Refining the test phase of usability evaluation: how many subjects is enough?, *Human Factors*, **34**(4), 457–68.

WHICH? (1996) Route planning made easy, July, p.39.

# PART THREE

# Guidance on and Examples of Product Design

# CHAPTER TEN

# Developing a qualitative sense

ALASTAIR S. MACDONALD

*Product Design Engineering, Glasgow School of Art*

## 10.1 Introduction

What makes one product more appealing than another, where each is equally reliable and offers a similar level of technical performance, within a given price range? Products are often instinctively perceived, and product choice is rarely just an exercise in logic. Apart from the satisfaction of fulfilling a utilitarian function, an object can give its user pleasure, not only in terms of its ergonomic 'fit' but also through its aesthetic qualities. A user's subjective response to a product is a complex phenomenon, but it is important to understand some of the subjective factors influencing product choice and acceptability.

There are indications that human factors professionals are becoming increasingly interested in understanding what it is in products which gives people pleasure (Jordan and Servaes, 1995; Fulton, 1993). Often an ergonomic feature has a tangible aesthetic quality. Designers have experience in 'fine-tuning' not only a product's aesthetic but also its ergonomic qualities, and here they rely on important information from the ergonomist.

There is no doubt that the language and techniques used to share information and ideas between the worlds of the human factors professional and the product designer could be improved, and the ambition of this chapter is to find some common language between the two professions. It discusses qualities which embrace both the ergonomic and the aesthetic, and describes a series of tools to help develop a common language for human factors professionals and designers with the aim of developing an understanding of qualitative values.

## 10.2  Sensorial Data

'To perceive things means becoming aware of things/experiencing things through our senses. If we are then talking about design and aesthetics, we are talking about sensorial experiences with man-made objects, design' (Rotte, 1993). A complex process determines the quality and value of that sensorial experience. At a physical level, Dreyfuss (1967) discusses an 'environmental comfort zone' of two layers. The first layer is described as a 'bearable zone limit'. Within the second, 'great discomfort or possible damage is encountered'. Limits for pressure, temperature, humidity, oxygen and light levels are clearly quantified. While this information is vital, it is very difficult for the designer to know what value it has in *human* terms: data in this form have some way to go before they have a qualitative value. This format for discussing and presenting data has traditionally been used with the aim of avoiding discomfort or injury rather than proposing a 'qualitative' threshold – one which would enhance a user's experience. This is an argument well developed by Fulton (1993) and the whole tone of her essay calls for a more holistic and positive view of human factors and for collaboration between human factors professionals and product designers. She says:

> The human factors profession was actually developed specifically to respond to problems and deficiencies – to errors, discomfort, injuries, delays and low productivity. The interest in understanding human physiological capabilities and limitations has been largely aimed at eradicating such problems ... But it underplays the potential value of human factors and design. The new challenge is to make products and environments which do better than keep us from getting sick, damaged or irritated. Products and environments can enhance the quality of our physiological experience. To achieve this, we need to explore new ways of learning about people in dynamic interaction with products and new ways of implementing what we find out.

To advance the argument for a more useful format for the presentation of information relating to the senses, one can find an exemplary model developed by Pirkl and Babic (1988) which translates quantitative data for the senses – vision, touch, hearing and movement – into design guidelines. Pirkl has been concerned with the area of transgenerational design – designing products which are 'barrier free', i.e. able to be used by a wide range of ages and abilities. The demographics of an aging Europe offers new political, social and design opportunities, and the whole area of designing for an aging population is increasingly being seen as an important area to address (Coleman, 1993; Macdonald, 1996a). However, there is a paucity of physiological data on the elderly which are in a useful form for designers. In Pirkl's model, specific sensory components, e.g. visual acuity (clarity); visual accommodation (focus); visual threshold; colour perception; hearing acuity (volume); directional hearing; tactile sensitivity; cutaneous factors; and motor movements, are listed

## Ageing Process

| Specific Component | Physical Change | Functional Effect | Result (Problem) |
|---|---|---|---|
| Visual accommodation (focus) | Loss of flexibility of lens. | Range of focus decreases. | Specific decrease in ability to focus on nearby objects or read fine print (presbyopia). |

80+ ———
60 ———
40 ———
20 ———
1 ———

## Industrial Design Response

| Design Guidelines | Design Strategies |
|---|---|
| Minimise the need for typography. | ■ Substitute simple unequivocal graphic symbols.<br>■ Use positioning to communicate importance, order, and relationships of components, controls, and operations.<br>■ Eliminate irrelevant information and decoration. |
| When typography is essential for information purposes, make it legible. | GENERAL<br>■ Use appropriate size and weight of type, and also letter, word, and line spacing.<br>■ Favour sans serif type for nontext material.<br>■ Use upper and lowercase type for maximum readability.<br>■ Maximise contrast between type and field.<br><br>SPECIFIC<br>■ Avoid ornate, decorative, competitive, or segmented type faces for information messages.<br>■ On dials, use only whole numbers with a minimum of markings.<br>■ Combine type with graphic symbols, when possible.<br>■ Determine size of type by level of illumination and viewing distance.<br>■ Isolate individual information messages. |

**Figure 10.1** One of Pirkl's demographic charts: vision (source: reprinted from Guidelines and Strategies for Designing Transgenerational Products: An Instructor's Manual, by James J. Pirkl and Anna L. Babic (Acton, Mass.: Copley Publishing Group, 1988), by permission of author. Copyright © 1996 by James J. Pirkl. All rights reserved.

and tabulated. Physical change, functional effect and result (problem), are cross-referred to the 'industrial design response' to give 'design guidelines' and 'design strategies' for each sensory component by broad age grouping and with reference to the aging process. For example, the aqueous fluid of the eyeball becomes progressively more yellow with age, altering a range of colour perceptions and contrasts. Pirkl's charts, as in Figure 10.1, allow the designer to understand the deterioration in sensitivity and give a better starting point for designing.

This translation of sensorial and physiological data into this type of format is a relatively straightforward process and one which would help both professions to understand each other's needs and offer greater opportunity for both groups to work more closely together.

## 10.3  A Concept of Aesthetics

What is perhaps more difficult to discuss is how sensorial data provides not only a physiological but also personal and cultural values. First it will be necessary to discuss the concept of aesthetics. Rotte (1993) distinguishes two definitions of aesthetic: firstly, formal aesthetics in the philosophical sense, and secondly, aesthetics learned or experienced in perception. The second is useful to our discussion. He says,

> If we use aesthetics in relation to perception and not in the philosophical senses
> as the study of beauty, then new terrain can be explored.

One can see in the early figure drawings of Albrecht Dürer, from his Dresden sketchbook of the early 1500s, for instance (Strauss, 1972), or more recently from the anthropometric charts of Henry Dreyfuss (1967) or PeopleSize (Friendly Systems, 1994), that the body can be measured and quantified in physical terms. This corresponds to the image we can see in the mirror reproducing our anatomical proportions: a geographical topography.

However, there is another way of mapping the body – in sensory terms. Picasso's *Weeping Woman* (1937) charts the feeling of pain and anguish. Penfield's diagram of the 'homunculus' (Blakemore, 1976) maps out the proportional representation of bodily parts on the surface of the right cerebral cortex (Figure 10.2). The mouth and the thumb, for instance, appear much larger than the leg. 'It is like an electoral map as opposed to a geographical one... reproducing its Parliamentary proportions' (Miller, 1978). This is an image of the self which we feel through our senses as we inhabit the world.

The original Greek *aisthetika* meant 'that which is perceptible through the senses'. Aesthetics can be a devalued term. It is concerned not just with visual form, colour or texture, but also with understanding and anticipating the effects of sensory stimulation on human perceptions and cognition. It is concerned with 'beauty', that combination of all the qualities of a person or object that

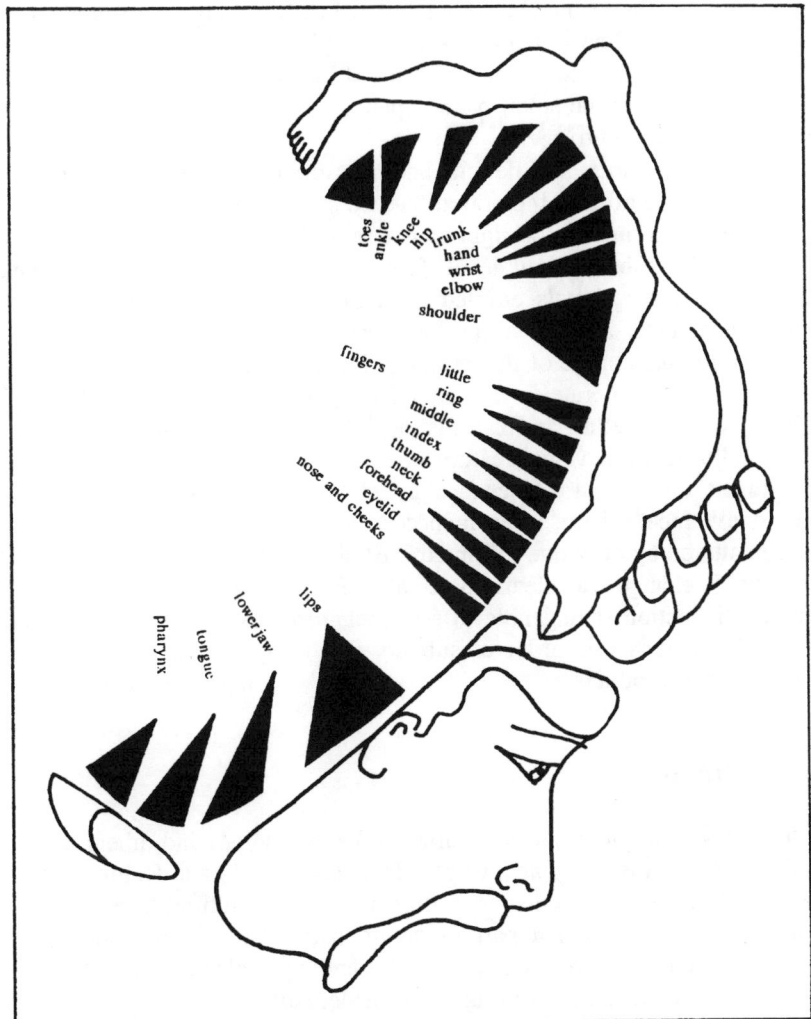

**Figure 10.2** Homunculus (after Penfield).

delight the senses and please the mind, and 'taste', the ability to make judgements according to a generally accepted standard.

All bodily faculties through which sensation is aroused – sight, touch, taste, smell, hearing, balance, movement and muscular effort – help to form an aesthetic appreciation of an object or environment. It is not only a sense of awe that one may feel at the extent of space and magnificence of architecture of a medieval cathedral like Chartres, but also an experience which embraces all our senses: our responses to light, colour, sound, smell, and temperature constitute a total aesthetic sensation (Macdonald, 1993a). Similarly, the sound of a fly fishing reel, as the line is being taken out, is like no other, and immediately taps into those pleasant personal memories of long summer evenings, the sound of flowing water and the smell of the open air – a complex aesthetic response evoked given the right triggers.

In products we look for delight and reassurance through our senses, for feedback which reinforces the notion that we have made the right decision, and that we are embarking on the right course of action. One can respond at both a conscious and a subconscious level. In successful products, designers seek to control the sensory signals emitted by a product – its appearance, texture, smell, temperature, as well as performance. Designers, amongst their other roles, have to take charge of this aesthetic language and should see themselves as communicators, literate in an non-verbal product language. Our senses are extremely discriminating and can distinguish subtle detail. We can detect the minimal adjustment of visual features. Seymour (1996) demonstrated just how finely tuned our visual senses are by showing three slides of a famous supermodel's face, where, in each successive photograph, the size of the nose had been altered by only one millimetre. By the third photograph the subtle but unmistakable change was clearly detectable. Each sense is as finely tuned, 'It's all about distinction, being able to see, feel, hear, experience the difference' (Rotte, 1993). It is this order of subtlety that designers have to control and embody in their products.

## 10.4  Culture

Are there aesthetic preferences peculiar to the individual, and others to groups of individuals or to society as a whole? Hofstede (1991) and Robinson (1993) offer useful models which discuss different layers, or cultural 'filters'. If social and cultural factors have a part in giving particular meaning and value to sensory experience, it would be useful to develop a clearer understanding of the concept of 'culture'. According to Hofstede, culture is,

> the collective programming of the mind which distinguishes the members of one group or category of people from another. Culture is learned, not inherited. It derives from one's social environment, not from one's genes. Culture should be distinguished from human nature on the one side, and from an individual's personality on the other.

Hofstede's model suggests three levels: those values specific to an individual; those specific to a group or category; and those which are universal, i.e. part of our human nature (Figure 10.3)

He goes further to discuss how cultural differences manifest themselves in different ways using an 'onion diagram' (Figure 10.4) to indicate how 'symbols represent the most superficial layer and values the deepest manifestations of

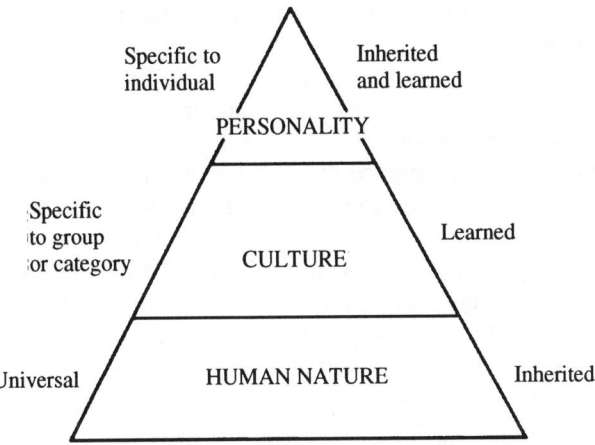

**Figure 10.3** Hofstede's three levels of uniqueness in human mental programming.

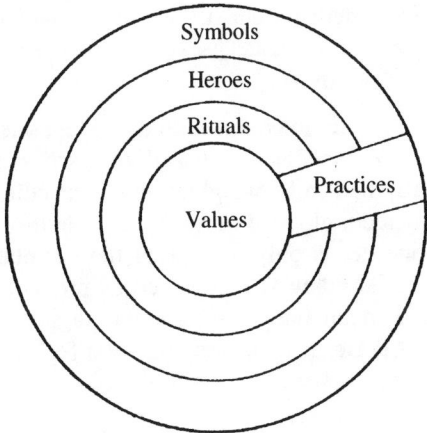

**Figure 10.4** Hofstede's 'onion diagram': manifestations of culture at different levels of depth.

culture with heroes [and here I would add icons] and rituals in between'. He describes values as 'broad tendencies to prefer certain states of affairs over others.' Sensory perceptions are given particular significance through social and cultural values. The quality of sound that a camera makes as the shutter closes, the satisfying clunk, or tinny clink a made by a car door closing, the smell of its interior, the radius on a pressed body panel indicating its thickness (implying quality) of metal, or the sound of its exhaust (this ploy has been used by the Japanese, recording the sound frequencies of exhausts of successful sports cars and then engineering for that effect), inform us of the 'supposed' quality of that product through our aesthetic intelligence, of the value that we place on these particular properties.

A complex perceptual, cognitive and cultural process allows us to make value judgements based on experience and preference. We betray cultural prejudices. To illustrate a point, take the example of a personal stereo: we can discriminate a cultural preference through touch and feel, by size and weight alone. A thin, heavy personal stereo might be judged superior in quality to a fat, light one, as these sensory signals correspond to our western cultural values of weight being associated with quality, and thin with high tech. Here, a rudimentary matrix (Figure 10.5) might begin to map out physical and sensory properties against cultural values for these sensations.

Manzini (1988) discusses at length the correlation between weight and value:

> The equation 'weight = importance' is not applicable to all cultures. A nomad, for instance, necessarily develops a system of light portable objects, and builds its meanings upon that system. Japanese history reveals, for that matter, a refined culture of lightness. In European culture, on the other hand, the value of weight (and the equations that derive from it, such as 'weight = quality, longevity, solidity, safety), is rarely rejected, and has certainly left its mark on the quality of our physical surroundings'. (p. 97)

In a similar way, early plastic injection-moulded telephones, though very light, did not conform to our expectations based on earlier and heavier bakelite handsets and eventually had to be weighted until our cultural values shifted to accommodate the new technological norm. Extending this idea into the field of automotive design, one needs only to look at the correlation between weight and perceived quality: Germany is recognised as producing only quality cars. The radius on pressed metal body panels often suggests a heavier gauge of metal than that used. All Germany's cars convey a feeling of solidity, of well-put-togetherness – part of Manzini's equation.

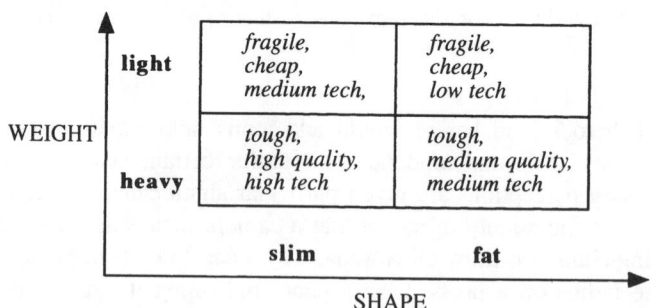

**Figure 10.5** Mapping out physical and sensory properties against cultural values for a personal stereo.

What emerges as a model is the idea of a flexible matrix which can map out the physical characteristics and associated qualities for a particular class of product. As one can map out the varying values associated with permutations of weight and a second factor, e.g. thickness, in a particular product, one could repeat this exercise for other combinations of sensory qualities and values in a range of products. There is some interesting work to be done here to reveal the 'cultural homunculus' within.

## 10.5 Understanding Users

Aesthetics is not a matter of absolute values – an understanding of context is vital. Visual symbolism is not universal. Perception, recognition and acceptance of an object is determined by the context in which it is used and by the nature and cultural conditioning of the user. Black is worn at funerals in this country, but in the East one wears white (Cumming and Porter, 1990). There are many aspects of colour which a designer has to handle. It has a physiological effect on the retina; it can be used in a structural way to differentiate elements of a product or affect the perception of its proportion, shape, etc.; the ergonomic function of colour assists the human/product interface to consider such issues as safety and visibility; and it has an aesthetic dimension which can reflect users' cultural preferences and emotional disposition.

Designers and human factors professionals need to be aware of the user's social and emotional needs as well as his or her physical and technological environment. With shifting values and trends, widening global markets, and more segmented niche markets, the product development team must attune to users' sensibilities. Different groups of people are on the look out for, and respond to, different signals. We are quite tribal in our allegiances, and have our own sets of preferences, each with its own visual codes and meanings, aspirations and values. The emerging field of sociological research known as 'user research' is becoming increasingly important in determining what people actually do and how they do it. According to James Woodhuysen of the Henley Centre for Forecasting, 'we must research users, watch them, listen to them, understand them ... this is going to be very important for the future of design'. Design practice is increasingly 'fanatically consumer orientated' (Baxter, 1995) and strategic design now requires its practitioners to 'focus on the user like a laser' (IDEO).

With demographic shifts in an aging population, as in Pirkl's example, the organising of data for another group of users is seen as increasingly urgent and, for big business, commercially valuable. The European Design for Ageing Network's pilot teaching pack 'Incorporating age-related issues into design courses' contains, in the context of human factors, a useful section on design strategies for older people (Chan, 1996) where the aim is 'to give students an overall view of different design strategies/methodologies experimented with

# a user research typology?
## 9 different groups of research methods....

**TYPOLOGY**

**FUTURE CREATOR**

* These people create and test future products and systems by living and working in their future world. Xerox parc is the exemplar of its field. The biggest problem is linking some of these far fetched concepts with more 'real-world' development centers (see parc again): many of the concepts have diminished transfer value.

**IMAGINE AND ACT OUT**

* These methods are a bridge from the real past to the imagined future. Using current general information, observations and trends, different scenarios of the future are created to 'practice the future'. The scenarios make the future real and understandable (and believable). Their value lies not in the accuracy of the 'predictions' but in the practice and flexibility you should gain.

**PROFESSIONAL TRACKERS**

* The predominant basis for this group is that trends can be identified and proven by finding many instances of that trend. So-called futurists like John Naisbitt and Alvin Toffler exemplify this group, providing useful general information, but the drawback is that any trend big enough to be spotted, is also obvious to many other people.

**DIRECT DESIGN EXPERIENCE**

* This is where designers typically live, developing things all the time, testing whatever product on themselves. It is an invaluable creative skill, allowing rapidly iterative prototyping skills, and one which should be prevalent through an entire organization. The drawback with this is the ownership of ideas: inventors love their creations, but love is blind (to sound cliched).

**CO-DESIGN**

* This entire category centers around designing (and all that entails) with end-users. The biggest difficulty is finding a communal space to undertake this, whether in a design studio, or in the user's natural environment. Software is leading the way here, with just about every software company using some sort of usability lab. The interesting thing about software is the mobility of the testing environment, as interfaces are 'virtual' but this omits some of the physical, natural surroundings vital for a productive workspace.

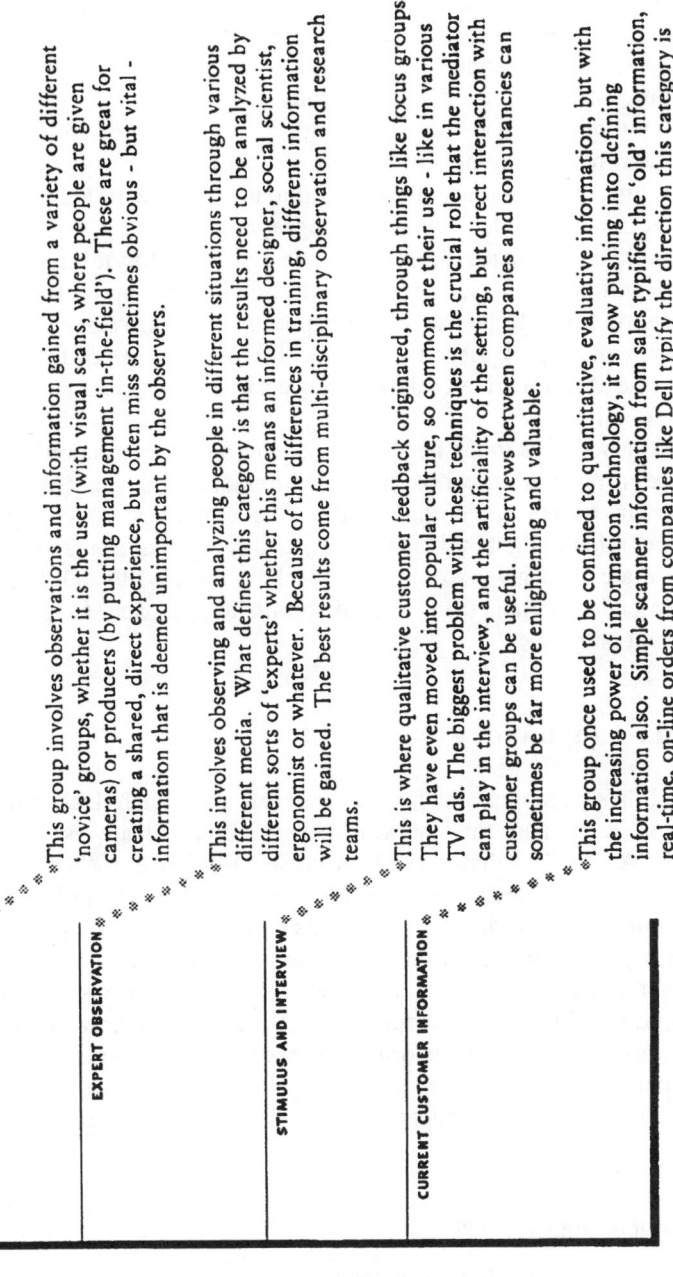

**CO-RESEARCH**

※ This group involves observations and information gained from a variety of different 'novice' groups, whether it is the user (with visual scans, where people are given cameras) or producers (by putting management 'in-the-field'). These are great for creating a shared, direct experience, but often miss sometimes obvious - but vital - information that is deemed unimportant by the observers.

**EXPERT OBSERVATION**

※ This involves observing and analyzing people in different situations through various different media. What defines this category is that the results need to be analyzed by different sorts of 'experts' whether this means an informed designer, social scientist, ergonomist or whatever. Because of the differences in training, different information will be gained. The best results come from multi-disciplinary observation and research teams.

**STIMULUS AND INTERVIEW**

※ This is where qualitative customer feedback originated, through things like focus groups. They have even moved into popular culture, so common are their use - like in various TV ads. The biggest problem with these techniques is the crucial role that the mediator can play in the interview, and the artificiality of the setting, but direct interaction with customer groups can be useful. Interviews between companies and consultancies can sometimes be far more enlightening and valuable.

**CURRENT CUSTOMER INFORMATION**

※ This group once used to be confined to quantitative, evaluative information, but with the increasing power of information technology, it is now pushing into defining information also. Simple scanner information from sales typifies the 'old' information, real-time, on-line orders from companies like Dell typify the direction this category is going.

**Figure 10.6**  Duncan's typology of user research methods (source: reproduced by permission, Moggridge, IDEO).

and adopted by leading practitioners in the field of designing for older people. Although older people are the particular focus, the strategies could be used for a broad range of sectors of society.

Accompanying this realisation is the steady growth and use of different user research methods. Duncan (1994) brought clarity and order to a diverse range of user research techniques by providing a useful typology of methods utilised by a range of companies and consultancies (Figure 10.6). Duncan's typology of methods includes video 'ethnography', usability testing, task analysis, empathising, scenario-building, design by storytelling (Moggridge, 1993) and trend forecasting, in addition to the more traditional designer's tools of lifestyle boards and scrapbooks, and provides examples of companies which employ these techniques. It is these types of tool which point the way ahead to creative collaboration between designers and human factors professionals.

Fulton (1993) also refers to the need for 'observation for inspiration' and the need to 'get under the skin' of people in order 'to understand what it feels like to be that person trying to do that work'. She continues, 'We begin to understand about what delights this person, what their special sensitivities and pleasures in life are' and gives examples of techniques used by IDEO which allow an open-ended rather than a predetermined approach to gathering information about people.

## 10.6  Product Language

One field which has generated much recent interest and which deals with the language of products and the messages they communicate is 'product semantics' (IDSA, 1984; Vihma, 1987; Aldersey-Williams, 1990). Product language can employ metaphor, allusion, and historical and cultural references. In innovative products, visual clues help to explain the proper use or function of a product and help minimise the need for text-based information. Whatever the academic debate surrounding various theories of product language or product semantics, it would be useful to discuss how the human factors, aesthetics and technical performance aspects come together to extend pleasurable operation for the user in a range of product examples. The ambition here is to discuss products using some of the concepts explored earlier.

### 1. Karrimor's Condor Rucsac Buckle

Karrimor's Condor Rucsac side-release buckle (Design Council, 1991) closes with a very positive 'click'. Visual, audio and tactile feedback are used so that it looks, feels and sounds good and ensures a reliable and reassuring fastening.

## 2. Good Grips® Kitchen Utensils

An example of 'universal design'. The handle is made from Santoprene®, a rubber-like thermoplastic which will not slip even when used with wet hands. The ribbing allows different handsizes to squeeze the handle into shapes comfortable to hold and use easily. The pleasure comes from these factors and the sympathetic tactile resilience of the material. The handle of the Samsonite Epsilon suitcase is made from a similar skin-friendly plastic material.

## 3. Global Knives

Global knives are a new concept in knives, designed and made in Japan. The blades are made from a molybdenum/vanadium stainless steel and are ice-tempered to a razor-sharp edge. The integral, hollow handles are weighted to give perfect cutting balance with minimum pressure required. Because of their smooth contours and seamless construction they allow no corners for food and germs to collect and they are exceptionally hygenic. Although consideration of the fit and comfort of the knife in the hand falls within the traditional preoccupations of human factors, the *sensation* and enjoyment of the finely balanced weight is an aesthetic one and one which can be sensitively tuned through the iterative and empirical process of making and testing used by the designer.

## 4. Zero Knives

The Zero knives by Seymour Powell are made from zirconium and have an extremely hard and sharp ceramic edge. However finely their fit, comfort or technical performance are improved, the *sensation* of cutting offered by these new materials is a pleasurable aesthetic one which the sensitivity and awareness of the designer has exploited.

## 5. Braun Micron-Plus Razor

The increasingly complex nature of new materials is normally described by their performance or qualities, e.g. a material which is endowed with a lightness, opacity, heat-resistance, elasticity, pliability, translucency, a special softness, a material which 'gives light' (an electroluminescent or photoluminescent surface), or materials which change form. Interesting use of material can be seen in the Braun Micron-Plus razor, designed in 1980 by Dieter Rams. Its body surface is made by overmoulding a soft elastomer thermoplastic polyurethane onto a polycarbonate body to give a decorative anti-slip and sound-deadening finish of a smooth surface with soft projecting points. The idea for these plastic studs was borrowed from ski-boot soles (Manzini, 1989, p. 198). The 'rubberised' coating is not only functional in providing grip, increasing the tactile pleasure in handling, but is also used for deadening the sound of the motor – a combination of the ergonomic, the aesthetic and the technical.

## 6. Neen Pain Xenos

The Neen Pain Xenos is a drug-free electronic pain relief device using electrical nerve stimulation. The Xenos works as a device which can be worn visibly as it refers visually through its size and proportions to the personal stereo – a socially acceptable product worn outwardly on the body. This has been made possible by using miniaturised electronic technology and touch controls which remove it from the alienating visual vocabulary associated with traditional medical equipment (Design Council, 1991).

## 7. NovoPen™

Those suffering from diabetes traditionally had to use clinical-looking syringes and needles. The NovoPen™ offers a much more acceptable model for the self-administration of precise amounts of insulin as its appearance now alludes to a much more (socially) acceptable product – the technical pen. This has a positive signal rather than the more negative one of the hypodermic syringe, now further coloured in its association with drug abuse. The NovoPen™ incorporates tactile and colour codes which refer to the different types of dosage of insulin. These provide sensory back-up and contribute to the aesthetic profile of the product. The quality of the sound of the 'click' as a dose is prepared for delivery is determined by the quality of the plastic chosen for a discrete but positive sound. The surface texture is achieved by spark-erosion of the mould to provide a pleasant tactile quality, easy to use (once learnt), reassuring and acceptable. The technicalities of administering precise dosages have been translated into easy steps. There is perhaps no need to debate where aesthetics stops and where human factors begins, but to develop the idea that sensory data have a range of values – comfort, fit, pleasure and cultural value and meaning. In this product one can observe the successful synthesis of the technical, aesthetic and human factors requirements.

## 8. Olivetti Office Machine Control Panels

King and Miranda's control panels for office machines display an understanding of the need for reassurance by employing feedback from more than one of the senses (Barbacetto, 1987). Here one can see the coming together of engineering, aesthetic design and ergonomics. Not only do the keyboards incorporate good ergonomic understanding, but they also develop the idea of pleasurable operation through the degree of tactile feedback in the buttons, the visual layout and colours used. King and Miranda discuss their process of 'sensory decomposition' and 'sensory recomposition': where, in a traditional control panel key, the finger operating it would normally obscure the information. They have separated the information aspect from the operational area and introduced a tactile dimension to the key which also uses finely tuned acoustic feedback. It is very difficult to separate out, in cases like this, the aesthetic from the ergonomic: just what amount of pressure on a computer key and what choice of materials not only contribute to ease of operation, but also give that pleasant clatter as you perform your task?

## 10.7 Conclusions

A small number of human factors professionals are beginning to explore this interesting boundary between the two fields of human factors and aesthetics. There is clearly a desire to articulate issues which have not historically been within their professional sphere. The objective nature of their traditional methods has not given them a language with which they feel at ease in discussing subjective values – in fact, they have had little language to explore and express such ideas.

Robinson (1993) has proposed extending the field of human factors beyond the usual physical/cognitive 'fit' between products and people to embrace social and cultural considerations, personal needs, desires and aesthetic responses. He develops the idea of 'fit' at personal, social and cultural levels, rather in the way Hofstede has.

Looking beyond definitions of usability and comfort, Jordan and Servaes (1995) have begun to categorise the emotional responses from a group of individuals towards a range of consumer products by using such terms as security/comfort, confidence, pride, excitement, satisfaction, entertainment, freedom, and sentiment/nostalgia. They have also considered a range of design properties associated with pleasurable products, and aesthetics is one of nine they list. A designer often has to bring together various issues (quantitative and qualitative) simultaneously in a product to embrace not only performance, fit, tolerance, comfort or pleasure but also meaning, value and delight.

Fulton (1993) suggests taking a more holistic and positive view of the field. She states that:

> the aesthetic vision will certainly be best realised if it is shared by the whole design development team... By linking [human factors] viewpoints up with the design profession – whose essence is application and change – we can use each particular discipline to broaden our knowledge of the human consequences of design decisions. At the same time, the designer's strength is to help direct specialist methods and attention to issues which have particular value and relevance.

There is great scope for collaboration between those concerned with human factors and those familiar with handling aesthetic qualities. One way to start this process of enhanced collaboration is to provide more opportunity for human factors and product design students to work together to develop a shared language which would not only benefit young designers, who clearly recognise the need for this field of expertise in their work, but also provide future ergonomists with the understanding of what designers need and want – in a format they can work with – and in the context of design practice (Macdonald, 1996b). To facilitate this collaboration, tools will be required. Some of these are already available and have to be used more widely. Others require to be developed, and still more require to be invented.

There is room within the field of human factors to develop a range of tools which help to discuss and embody more of the aesthetic dimensions of a product, to take acount of the sensorial and the cultural, as well as the physical and cognitive – to develop that qualitative sense.

## References

ALDERSEY-WILLIAMS, H. (1990). *Cranbrook Design: The New Discourse.* (New York: Rizzoli.

BAXTER, M. (1995) *Product design.* London: Chapman & Hall, chapters 4 and 6.

BARBACETTO, G. (1987) *Design Interface: How Man and Machine Communicate.* Olivetti Design Research by King and Miranda. Milan: Arcadia Edizioni, p. 81.

BLAKEMORE, C. (1976) *Mechanics of the Mind,* BBC Reith Lectures 1976. Cambridge: Cambridge University Press, pp. 79–80 (illustration redrawn for Miller, 1978, p. 21).

CHAN, S. (1996) Design strategies for older people. In: Hewer, S. (ed.), *Design Age Pilot Teaching Pack.* London: Royal College of Art.

COLEMAN, R. (1993) *Designing for Our Future Selves.* London: Royal College of Art.

CUMMING, R. and PORTER, T. (1990) *The Colour Eye.* London: BBC Books.

DESIGN COUNCIL (1991) *British Design Council Awards.* London: Design Council.

DESIGN MANAGEMENT INSTITUTE (1993) *Novo Nordisk A/S: Designing for Diabetics.* Boston: DMI Press.

DREYFUSS, H. (1967) *The Measure of Man – Human Factors in design.* New York: Whitney Library of Design.

DUNCAN, E. (1994) User Research, IDEO draft document for discussion purposes. San Francisco: IDEO.

FRIENDLY SYSTEMS (1994) *PeopleSize: Intelligent Computer Database of Human Dimensions.* Loughborough: Friendly Systems.

FULTON, J. (1993) Physiology and design new human factors, *American Center for Design Journal,* 7(1).

HEWER, S. ed. (1996) *Incorporating Age-related Issues into Design Courses,* Design Age Pilot Teaching Pack. London: Royal College of Art.

HOFSTEDE, G. (1991) *Cultures and Organisations.* Maidenhead: McGraw-Hill International.

IDSA (Industrial Designers Society of America) (1994) The semantics of form, *Innovation Journal,* Spring.

JORDAN, P. and SERVAES, M. (1995) Pleasure in product use: beyond usability, *Contemporary Ergonomics 1995*, pp. 341–6.

MACDONALD, A. S. (1992) *Aesthetics in Engineering Design*. Loughborough: SEED.

MACDONALD, A. S. (1993a) Developing a qualitative sense. In: *Aesthetics in Design: Colloquium, Institution of Electrical Engineers, Digest No. 1993/153, 29 June 1993* pp. 5/1–5/4. London: IEE.

MACDONALD, A. S. (1993b) *Ergonomics in Engineering Design*. Loughborough: SEED.

MACDONALD, A. S. (1996a) Participating in the process, *The Challenge of Age*. Glasgow: Glasgow School of Art, pp. 69–78.

MACDONALD, A. (1996b) *User-friendly Human Factors for Design Students*. Paper submitted for review.

MANZINI, E. (1988) *The Material of Invention: Materials and Design*. London: Design Council.

MILLER, J. (1978) *The Body in Question*. London: Jonathan Cape, p. 21.

MOGGRIDGE, B. (1993) Design by storytelling, *Applied Ergonomics*, 24(1).

PIRKL, J. J. (1994) *Transgenerational Design: Products for an Aging Population*. New York: Van Nostrand Reinhold.

PIRKL, J. J. and BABIC, A. L. (1988) *Guidelines and Strategies for Designing Transgenerational Products: A Resource Manual for Industrial Design Professionals* and *Guidelines and Strategies for Designing Transgenerational Products: An Instructor's Manual*. Acton, Mass.: Copley Publishing Group.*

ROBINSON, R. (1993) 'What to do with a human factor', *American Center for Design Journal*, 7(1).

ROTTE, A. (1993) Design and aesthetics – ordo, claritas et consonantia. In: *Aesthetics in Design: Colloquium, Institution of Electrical Engineers*, Digest No. 1993/153, 29 June 1993, pp. 3/1–3/6. London: IEE.

SEYMOUR, R. (1996) Business masterclass, Glasgow International Festival of Design.

STRAUSS, W. L. (ed.) (1972) *Albrecht Dürer: The Human Figure. The Complete Dresden Sketchbook*. New York: Dover Publications.

VIHMA, S. (1987) (ed.) *Form and Vision*. Helsinki: University of Industrial Arts.

* The transgenerational design guidelines and strategies charts are reproduced from *Guidelines and Strategies for Designing Transgenerational Products: A Resource Manual for Industrial Design Professionals* and *Guidelines and Strategies for Designing Transgenerational Products: An Instructor's Manual* by James J. Pirkl and Anna L. Babic, published in 1988 by Copley Publishing Group, Acton, Mass., USA. The project was supported, in part, by grant no. 90-AT-0182, from the Administration Office of Human Development Services, Department of Health and Human Services, Washington, DC 20201; and the Gerontology Center and the Center for Instructional Development of Syracuse University, Syracuse, NY, USA.

# CHAPTER ELEVEN

# Applying ergonomics methods during the industrial design of consumer products

MARK EVANS

*Department of Design and Technology, Loughborough University*

## 11.1 Introduction

In attempting to achieve a competitive advantage a reduction in time to market offers substantial advantages to manufacturers. By compressing product design timescales, overall costs can be reduced and products launched ahead of those of competitors. Unfortunately, without careful management, reduced leadtimes can also result in substandard products, in terms of both manufacture and design. This is becoming all too apparent as product recall notices appear in the national press with increasing frequency. The reasons for product recall are not restricted to material and production deficiencies. The absence of ergonomic evaluation can result in design-centred product recalls.

From the experiences of the author as both an in-house and consultant industrial designer, there appears to be a lack of support for the use of ergonomic evaluation during the product development process. The reasons for this being largely due to time, cost, and a lack of awareness as to what ergonomics can contribute to the product.

This case study for the industrial design of a nylon-line garden trimmer was supported by the application of ergonomics methods. It identifies a methodology whereby ergonomics and industrial design can be integrated into programmes of new product development (NPD). Whilst the result of these activities added to the overall product development time at the start of the programme, this was offset by the use of computer aided design (CAD) and rapid prototyping. In addition to this, the design output could be demonstrated as being a quantitative improvement on products currently available.

## 11.2 Evaluation Programme

In order to identify existing problems and user perceptions of line trimmers, an evaluation programme was undertaken. This involved a market survey the use of the home accidents surveillance ID system (HASS), a general user questionnaire, and detailed user trials.

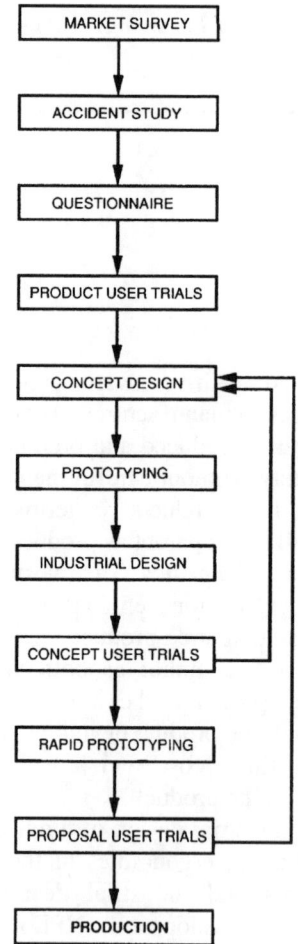

**Figure 11.1**  Schematic diagram of the design and evaluation programme.

The findings from these activities were then used to produce a concept design which was subjected to further ergonomics evaluation via the use of models, prototypes and computer renderings. This critique of the concept design allowed further development leading to a more refined design proposal. A schematic diagram of the overall programme can be seen in Figure 11.1.

## 11.3 Results

### 11.3.1 Market Survey

Whilst industrial designers generally engage in some form of market survey, their methods tend to be less rigorous than those of ergonomists. Industrial designers are more inclined to focus on areas that concern them during their design activity, namely appearance, manufacturing techniques and the overall 'feel' of the product.

During the case study, one of the ergonomics evaluations involved determining the centre of gravity of existing products. In every product tested, the centre of gravity was some distance from the hand-grip.

### 11.3.2 Questionnaire

Two hundred questionnaires were posted, 76 being returned completed. Findings from the questionnaire indicated the extensive use of products with a lawn-edging capability (41 per cent ownership), and that 94 per cent of the owners found this feature useful.

Possibly as a result of the poor balance identified during the market survey, the presence of an extra handle was considered useful. Whilst all of the respondents found it useful to have a secondary handle for extra support, further analysis revealed this to be a reaction to the centre of gravity being so far away from the main handle. The secondary handle had little to contribute to the control of the product, it simply made it easier to lift.

### 11.3.3 Accident Survey

Whilst the HASS data were available free of charge, they were not specific enough to be of significant value to the design programme. Indeed, some of the reports tended to have a tenuous association with line trimmers such as, 'whilst playing brother hit him with trimmer'.

It would have been useful to follow up incidents with questionnaires but this would have required excessive time and resources.

### 11.3.4 User Trials

Four electric line trimmers were identified which had all of the key features characteristic of products currently on the market, i.e. wrist support, secondary handle, battery powered, and heavy duty motor. Using a set procedure, sixteen users undertook observed trials and were questioned on their reactions to each product (Figure 11.2 shows a trial in progress).

Neither the wrist support nor the handle that had no bend was liked. Another relevant finding was that whilst the results of the questionnaire identified that a secondary handle was useful, there was general dissatisfaction with the comfort afforded by such handles.

The fact that the centre of gravity was away from the handle became an area of interest during the market survey. It was therefore not surprising to note a general dissatisfaction with balance. The only product that received a significant positive feedback was the machine on which the battery was positioned behind the handle, thereby shifting the centre of gravity towards the handle. Apart from the battery machine, all were considered generally poor to use one-handed. This was unfortunate as, during use, one-handed operation allowed the greatest area to be cut.

Not surprisingly during two-handed operation using the extra handle there was a major reduction in effort and corresponding increase in comfort. The only problem with this was the resulting loss of cutting range owing to the restricted movement.

**Figure 11.2** A user undertaking evaluation of an existing design.

The presence of power cables was highlighted as a negative feature by the majority of users. The cables made use more difficult, and there was perceived to be a safety issue associated with the line cutter slicing the flex. In contrast to this, the battery machine had no cables and removed the negative features associated with them.

## 11.4  Industrial Design

The results of the initial ergonomic studies were collated to provide a specification for the concept design. The features that appeared to be important were a lawn-edging capability and a bend in the stem just in front of the handle.

The notion of comfort was considered to be a function of product balance, and with a centre of gravity at the handle there should be a dramatic improvement in balance and therefore comfort by reducing the effort. There was also the potential for accessibility to the greater cutting area that is possible with one-handed use.

The techniques used for the development of a concept design are described below.

### 11.4.1  Sketching

The design programme started with a conventional approach whereby sketching was used to quickly generate and evaluate ideas on paper. Despite the fact that extensive use of CAD was planned, there is at present no substitute for the spontaneity and flexibility of sketching at the concept stage.

### 11.4.2  Foam Modelling

As a 2D medium sketching is limited in its ability to represent 3D products. In order to gain confidence in the ergonomics of the handle shape, a full-size foam model was produced. Whilst there are problems associated with industrial designers designing for themselves, the foam model was modified on an on-going basis until it represented the dimensions and requirements highlighted during the ergonomics evaluation. The foam model made a significant contribution to the concept user trials as it was possible to confirm the suitability of the handle dimensions.

### 11.4.3  Prototyping

From the ergonomic evaluations, the results on comfort, balance and range of use were interpreted as a requirement to position the centre of gravity at the handle. In order to achieve this without the addition of a counterweight, it was necessary to locate the motor behind the handle and to use a flexible drive to take the rotary motion to the cutter.

A working prototype was produced that included the key features of the concept design. This was essential for engineering evaluation and for subsequent concept user trials. Whilst it did not have the appearance of the intended product, its weight, performance, and balance were representative.

### 11.4.4   Computer Modelling

Having devised a basic component configuration, appearance and size, the design was transferred onto a CAD system for further detailing and the generation of a 3D computer model. The latter was necessary for the rapid prototyping which was to be used to save time on the production of physical models.

Having confirmed that the components would fit into the surfaces generated on the CAD system, the design was finally rendered to produce a rendering of the proposal. To add an impression of scale, lawn stripes were used as a background. These renderings were used during the concept user trials.

The final views printed out were of the cutter, handle and a general view of the entire product with its S-shaped stem. The cutter and handle details can be seen in Figure 11.3.

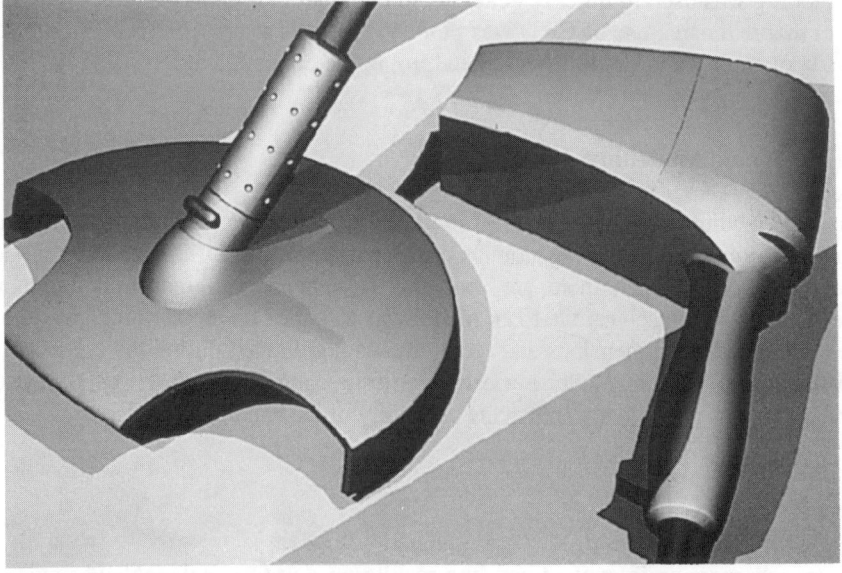

**Figure 11.3**   Computer rendering of the concept design.

## 11.5  Concept User Trials

As outputs of the industrial design phase, the working prototype, foam model, and computer rendering enabled the key features of the proposal to be evaluated. Whilst these represented three distinct model types, the key elements of the design were present, i.e. appearance, feel and performance. It was therefore possible to introduce these representations into a series of concept user trials.

Nine members from the product user trials were invited to comment on the new design following a refamiliarisation with the four products used. After familiarisation, the prototype was used to perform the same task as in the product user trials. On completion, the users were questioned on the key features present in the prototype, such as handle orientation, length and balance. The renderings were then used to evaluate appearance, and the foam handle to evaluate feel and overall dimensions.

Without exception the major features of the prototype received a positive response. In terms of appearance as communicated by the computer renderings, responses were less positive, with the trend being towards good and average. The foam model was well liked in terms of its feel, with all users preferring not to change any dimensions.

By asking, too, for general comments, a mismatch in the orientation of the handle on the computer renderings and prototype appeared. The angle on the renderings was steeper and pointing more towards the ground. This was not liked, and cross-referencing with the product user trials showed the same result for the existing product with a similar configuration. Modifications were therefore necessary.

## 11.6  Rapid Prototyping

During a more conventional industrial design programme, technical drawings would be prepared to enable the production of a visual model. Visual models are non-working representations that take considerable time (and cost) to produce. The main function of such models is to gain approval of the product appearance by senior management. Any ergonomics evaluation is extremely limited as visual models do not work, are easily damaged by handling, and, as they contain no components, are rarely of the correct balance.

A higher level of model is the visual prototype which is a fully working, handmade product that looks like the production item. These are relatively expensive to produce, take considerable time and require technical drawings for each component.

As the line trimmer was modelled on a CAD system, there was the potential to remove the need for a visual model and visual prototype by using rapid

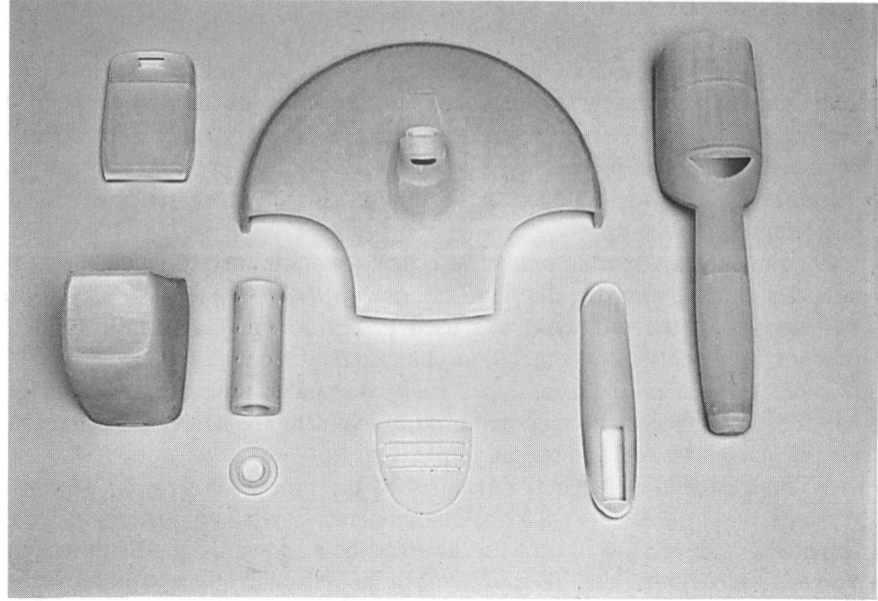

**Figure 11.4**   Rapid prototype components prior to assembly.

prototyping. Rapid prototyping uses a laser either to cut, to harden or to fuse material, therefore removing the need for technical drawings and complicated handmade components.

The mouldings for the line trimmer were produced using stereolithography, which uses a UV laser to harden layers of a photocurable resin. The components were produced in 36 hours (Figure 11.4), and required a further two working days for the inclusion of components and finishing. By using rapid prototyping, the visual prototype was produced sooner in the NPD programme, and at around 50 per cent of the cost of conventional techniques. It also removed the need for a visual model.

## 11.7  Proposal User Trials

Despite the use of rapid prototyping, the visual prototype was still relatively fragile, and by the end of the final trials the paint finish was chipped and quite dirty in places (made worse by the use of a matt finish). However, the structural integrity of the product was maintained, and the users were able to carry out an evaluation and approve the design.

The completed visual prototype produced using rapid prototyping is shown in Figure 11.5 before the proposal user trials.

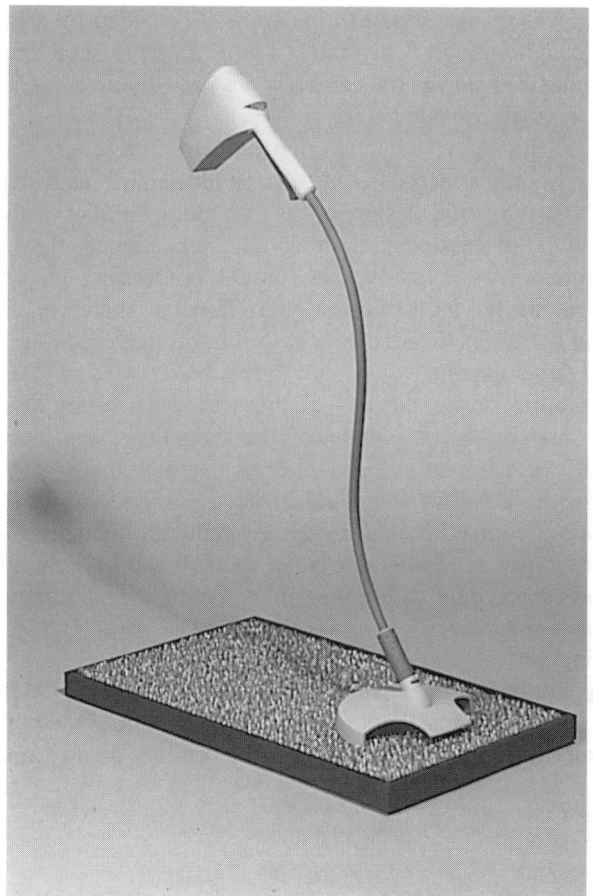

**Figure 11.5**   Visual prototype produced using stereolithography.

## 11.8  Conclusions

It is impossible to predict how the design might have evolved without the use of ergonomics methods. Industrial design is a creative activity, and innovation does result from professional practice. However, during this research, the application of ergonomics methods not only assisted in the definition of dimensions and product opportunities but also enabled benefits to the user to be represented in a quantitative format. The latter could be extremely useful in convincing a client of the need to move to a particular product configuration, especially if the client has been discouraged by increased costs.

One of the aims of this investigation was to identify how ergonomics methods could be applied by industrial designers. It is felt that the success of the programme was owing in part to the support provided by ergonomists and resources at Loughborough University. Quite clearly, such back-up is not available to industrial designers generally as they operate predominantly as individuals or in small consultancies.

It would be all too easy to declare that industrial designers should become proficient in ergonomics methods, but this would require considerable further education. Whilst industrial designers may become familiar with ergonomics methods, without professional training and experience they cannot call themselves ergonomists. Equally, one would not expect an ergonomist to undertake the role of the industrial designer. There is, therefore, a real need to make industrial designers aware of the contribution that ergonomists can make during product development.

During the product design process, clients will remain wary of any increases in timescales and/or costs, whatever the benefits. This may make the introduction of ergonomics to those organisations that most need them extremely difficult. Equally, some industrial designers will use advances in CAD to provide a competitive advantage by reducing their leadtimes further. Whilst both responses are understandable, there is the opportunity to employ CAD and rapid prototyping with ergonomics methods whilst maintain existing design timescales and costs.

The combination of the more subjective (industrial design) with the objective (ergonomics) is a powerful combination during product development. Closer collaboration will not only be to the benefit of the two professions and clients but also ultimately improve the safety, quality and appeal of manufactured goods.

CHAPTER TWELVE

# Design of hand-operated devices

ALAN HEDGE

*Dept. of Design and Environmental Analysis, Cornell University, USA*

## 12.1 Introduction

This chapter describes ergonomic principles for the design of hand-operated devices. A hand-operated device is defined as any product that is used as a hand-held product, such as a hand-tool, or is a hand-operated product, such as a computer keyboard or mouse. The chapter does not describe design information for product components, such as the shape, size or force required to operate any product controls, such as buttons or switches, or provide detailed information on the design of any visual displays on the product, such as liquid crystal displays or product labelling as might be found on cellular telephones, hand-held calculators or remote control devices. Rather, the focus of this chapter is on the design principles that allow any hand-operated product to be used in a good posture that enables the user to sustain productive work in a safe manner.

## 12.2 Hand-operated Products are Important

How many hand-operated products do you use daily? If you pause for a moment to think about the products that you interact with on a daily basis you will quickly realise why the good ergonomic design of hand-operated devices is important. Chances are that, on a typical day, you will brush your teeth with a toothbrush, eat with a knife, fork and spoon, drink from a cup, write with a pen, hold a telephone, turn a key in a lock, open a door using a handle, drive a car using the steering wheel, type on a computer keyboard, change channels using a TV remote control, cut something using scissors, and perhaps you will also use hand-tools like a screwdriver, power drill, shovel and so forth. In fact, it is hard to imagine life without hand-operated devices, but because they are so ubiquitous it is easy to overlook the need to pay attention to applying good ergonomic design principles to their design.

Hand-operated devices are ubiquitous and they are integral to our daily routines. Sperling *et al.* (1993) propose a useful framework for the work environment variables that affect the design and operation of a hand-tool, and this framework can be generalised to any hand-operated device (see Figure 12.1). The framework identifies four sets of variables that define the context for the use of any hand-operated device. All of these variables need to be considered in the design of any hand-operated device. They are:

- *User characteristics*   Characteristics such as the person's age, gender, body dimensions and training, because these affect hand size, strength and dexterity. A product such as a remote control may work well for a nimble-fingered youth but not for an elderly adult with arthritis.

- *Workplace*   The work process, the location of work (e.g. office, car, airport, on the street, etc.), the physical layout of the workplace and the prevailing climate conditions (e.g. cold or hot, wet or dry, quiet or noisy, bright or dim, clean air or dusty air, stable or vibrating setting) are considerations that affect how a hand-operated device is used. These variables will in turn influence the person's posture, especially wrist posture, and the force and precision that needs to be used to operate the product. For example, writing a letter with either pen and paper or a laptop computer is easy when in a quiet office, but much more difficult when on a bus on a rough road or in an airplane that is passing through air turbulence.

- *Work organisation*   The types of task, the frequency of work cycles, the quality of the work that is required and psychosocial factors all affect physical and mental stress, which in turn affects how well a hand operated device is used.

- *Hand-operated device*   The kind of hand-operated device and its physical design (size, weight, shape) determine how easily, safely and accurately the hand operated device can be used.

Sperling *et al.* (1993) also describe a useful cube model for hand-tools that categorises the tools in terms of three dimensions:

- *Time*   How frequently the hand-tool needs to be used.
- *Force*   Whether a low or high degree of force needs to be applied to operate the tool.
- *Precision*   Whether a low or high degree of precision is needed when using the tool.

This kind of model is useful for organising the relationships between these three dimensions and various hand-tool designs. It can also be generalised to other types of hand-operated device.

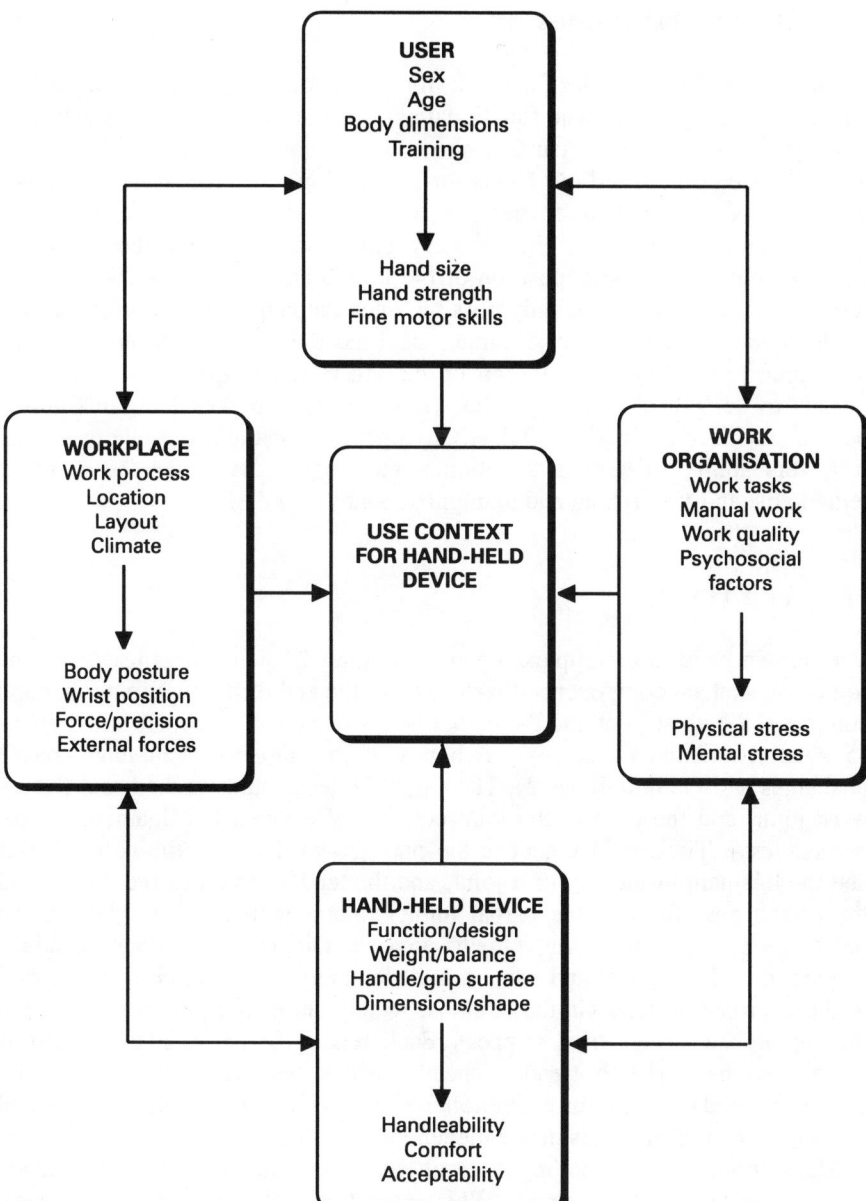

**Figure 12.1**  Design considerations that affect the use context for any hand-operated device (source: adapted from Sperling *et al.*, 1993).

## 12.3 Design Principles

Good ergonomics principles for the design of hand-operated devices depend on an understanding of the basic functional anatomy of the hand. The human hand is a dexterous grasping organ that allows us to manipulate objects in diverse ways. As we do not need our hands for locomotion, our bipedal gait has freed the hands, enabling them to perform a variety of tasks using a variety of implements we call tools. Evolutionary biologists have described how the development of the true opposition of the thumb and the fingers (prehensile), found only in humans, was only possible once the hand was no longer needed for locomotion. The prehensile human hand has the ability to bring the thumb into opposition with the distal joints of the fingers (fingertips) thereby allowing objects to be 'pinched' and gripped. Other primates do not have this ability because the thumb is not anatomically in a position that allows true opposition. It is this ability to oppose the thumb and fingers that gives the hand its remarkable abilities to hold and manipulate many tool designs.

## 12.4 The Hand

The human hand is a complex organ containing 27 bones organised into the following skeletal components. Starting from the end of the forelimb, the hand comprises the wrist joint and the carpal bones (8 bones), the metacarpal bones (5 bones), and the phalanges (14 bones of the digits – each finger has 3 phalanges and the thumb has 2). The carpal bones form an arched roof for the wrist joint, and the base of the joint consists of a tough flat ligament (flexor retinaculum). The carpal tunnel is the passageway between the carpal bones and the ligament inside the wrist joint, and the tendons that flex the fingers and the thumb pass through the carpal tunnel, along with a nerve (the median nerve). All major finger movements involve the use of tendons. Tendons connect muscle to bone, and when a muscle contracts, the force is transferred to the appropriate bone via the tendons, which causes the joint to move (like pulling on the strings of a puppet). Each tendon in the hand and wrist is surrounded by a sheath (tendon sheath) which secretes a lubricating fluid (synovial fluid) to minimise frictional forces as the fingers are flexed and extended (the tendon slides inside the tendon sheath).

Major movements of the fingers result from movements of tendons attached to the muscles of the forearm. When muscles at the back of the forearm contract, the extensor tendons open (extend) the fingers, and as contraction continues, the hand is pulled upwards from horizontal (this is called wrist extension). When the muscles at the front of your forearm contract, the flexor tendons close (flex) the fingers, and if this continues the whole hand bends downwards (wrist flexion). Tendon movement from full flexion to full extension can be 5 cm. Note that the flexor muscles of the forearm are much

**Table 12.1** Anthropometric dimensions of the hand (cm)

| | Men | | | Women | | |
|---|---|---|---|---|---|---|
| | 5th %ile | 50th %ile | 95th %ile | 5th %ile | 50th %ile | 95th %ile |
| Hand breadth (metacarpal) | 7.9 | 8.6 | 9.7 | 6.9 | 7.6 | 8.6 |
| Hand thickness (metacarpal) | 2.8 | 3.0 | 3.3 | 2.0 | 2.5 | 2.8 |
| Hand breadth (thumb) | 9.4 | 10.4 | 11.2 | 8.1 | 9.1 | 10.2 |
| Hand length | 17.8 | 19.3 | 20.8 | 16.3 | 17.5 | 18.8 |

*Source:* Adapted from Lewis and Narayan, 1993

larger than the extensor muscles, which confirms that the hands are developed primarily to flex the fingers in order to grasp objects.

In addition to major hand and finger movements, the fingers and thumb are capable of many finer movements, such as touching each finger tip with your thumb, wiggling your thumb around in a circular motion (this requires up to six muscles), crossing your fingers and so on. The hand contains many smaller muscles. Curling (flexing) or extending your fingers without moving your hand into wrist flexion or extension occurs using the small lumbrical muscles inside of the hand. Turn your hand palm up and you will see a fleshy ridge at the base of your thumb (the thenar eminence) produced by the muscles that allow you to flex and fold your thumb across the palm of your hand. At the base of your little finger (fourth digit) you will see another lesser fleshy ridge (the hypothenar eminence) produced by the muscles of the little finger. As you rest your arm and hand on a flat surface, notice the areas that come into contact with that surface, namely the digits, the hypothenar and thenar eminences, and the front of the forearm. The centre of your palm and your wrist crease both are arched away from the surface for good reason: both contain nerves and blood vessels close to the surface. This fact alone has enormous significance for the design of hand-operated devices because in shaping the device we have to ensure that it never puts pressure on these sensitive regions of the hand. Relevant design dimensions for the human hand are presented in Table 12.1.

## 12.5  Neutral Hand Posture

For the purposes of ergonomic design, a neutral hand posture is one where the hand is straight, not bent up or down or sideways or twisted. While the hand is in a neutral posture the fingers can be flexed or extended. For example, a well designed power drill can be held with a power grip with the hand straight.

Neutral posture is not synonymous with static posture, and we can work with the hands in neutral if these are moving within a neutral zone of motion. Working with the hands in neutral is a means of minimising the risk of a cumulative trauma injury to the hand or wrist. Research shows that hand movements affect interstitial fluid pressure within the carpal tunnel, and any pressure increase can compress the median nerve and other structures. Pressure changes within the carpal tunnel show a curvilinear relationship between vertical extension/flexion hand movements and carpal tunnel pressure (CTP) increases (Rempel *et al.*, 1994; Rempel and Horie, 1994). Extreme hand positions, such as acute flexion combined with ulnar deviation, can prevent the free flow of blood and other fluids into the palm of the hand, whereas these flow freely into the palm with the hands in neutral to moderate extension (less than 20°) or flexion (less than 20°: Gelberman *et al.*, 1984). With the hand in a neutral posture, CTP typically remains below 30 mmHg. Sustained increases in CTP above 30 mmHg are undesirable because they may detrimentally affect functioning of the median nerve. Animal studies shows that short-term increases in CTP, between 30 and 50 mmHg, can disrupt blood flow to the median nerve (Dahlin *et al.*, 1991). Studies in animals and humans show that nerve conduction velocity changes will occur when the CTP is in the range of 40–50 mmHg; sustained CTP in the range of 40–50 mmHg for an 8 hour period can result in complete blocking of nerve signals (Hargens *et al.*, 1979).

The range of wrist movement between 20° extension and 20° flexion is called the sector of maximal utility, and in this range there is minimal pressure on the articular surfaces in the wrist and the ligaments remain slack (Kapandji, 1982). CTP also exceeds 40 mmHg whenever the wrists are flexed or extended beyond 20° (Gelberman *et al.*, 1984). The lowest CTP occurs in the range of 0–15° wrist extension (Rempel *et al.*, 1994; Rempel and Horie, 1994). Palmar flexion of the wrist causes rotation of the anterior part of the lunate and its palmar projection, which compresses the median nerve and reduces the volume

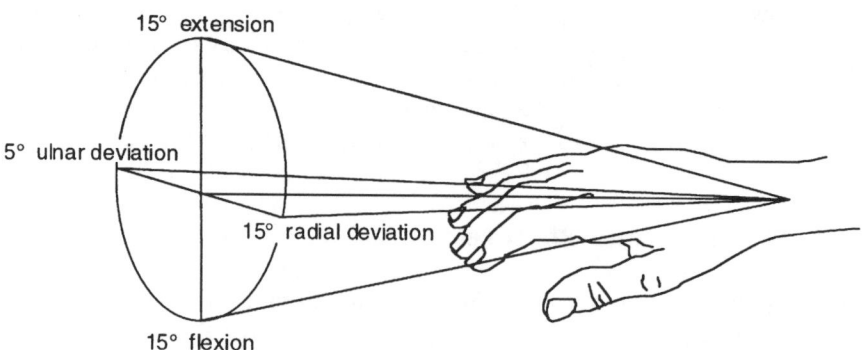

**Figure 12.2** A neutral zone of hand movement.

of the carpal tunnel (Robbins, 1963). Flexion of the wrist beyond 20°–40° increases CTP beyond 40 mmHg (Gelberman *et al.*, 1984). When the hand remains vertically neutral but moves laterally from side to side, comparable changes in CTP are not seen until the hand is in ulnar deviation in excess of 25° or radial deviation in excess of around 15° (Rempel *et al.*, 1994). Thus, a neutral zone of hand movement can be defined which is bounded by 15–20° wrist extension, 20–40° flexion, 20–25° ulnar deviation, and 15° radial deviation. Hand movements within this zone should produce minimal adverse changes in CTP and pressure on articular surfaces during hand activity. Good ergonomic design of any hand-operated device should allow the user to work comfortably while his or her hands are moving in this neutral range of motion (see Figure 12.2).

## 12.6  Holding a Hand-operated Device: The Five Basic Types of Hand Grip

The use of hand-operated products often requires us to hold or grip the product in some way. The hand postures required to do this can be categorised into five basic types of grip (Eastman Kodak Company, 1986), each of which creates different product design considerations.

### 12.6.1  Precision (Pinch) Grip

The pinch grip may have evolved to allow us to pick small objects from the ground, and this grip allows us to grasp small objects. The grip also allows us to use small hand tools, such as pens or sewing needles. It allows us to rotate small objects between the pulp of the thumb and fingertips, so we can operate small rotary controls, and it allows us to grasp small lever controls, such as switches, and to push and pull small objects, such as putting a disk into a computer or retrieving a coin from a vending machine. The pinch grip affords us a very high degree of precision control and hence is sometimes called a precision grip.

Opposition between the thumb and fingers is also maintained when we grasp objects to pick them up, such as lifting a telephone receiver or picking up a book. This type of grip can also be used to help us to stabilise the body, such as happens when grasping a deep rectangular hand rail. Other variants of this grip include the lateral grip that is used when we hold, say, a newspaper between the thumb and fingers. Sometimes a pinch grip is used to carry objects, such as carrying a dining tray in a cafeteria. We also use a pinch grip with both hands when we perform tasks such as screwing together a nut and bolt by hand. Strength tests show that, at best, the pinch grip can generate about 20–25 per cent of the force that can be generated by a power grip (see

**Table 12.2**  Optimum and maximum spans for a pinch
grip for a 50/50 mixed workforce (cm)

|           | 5th %ile | 50th %ile | 95th %ile |
| --------- | -------- | --------- | --------- |
| Optimum   | 2.1      | 4.3       | 7.9       |
| Maximum   | 10.8     | 12.5      | 15.0      |

*Source:* Eastman Kodak Co., 1983

below), and the distance between the thumb and fingers can dramatically
affect this. There is an optimum distance for a pinch grip of 2.5–7.5 cm: as
the grip distance gets less than 2.5 cm or more than 7.5 cm so the grip strength
decreases. However, the optimum grip span will depend on the type of task,
for example writing becomes more difficult with either a very thin or a very
fat pen. The ranges of functional hand grasp for a pinch grip are show in
Table 12.2.

Konz (1990) distinguishes two forms of the precision grip:

- *Internal precision grip*  Characterized by a pinch grip where the grip is
  supported by the little finger or the side of the hand, and the handle of the
  tool is 'internal' to the hand (e.g. the way that a knife is held – Figure 12.3).

- *External precision grip*  Characterized by a pinch grip where the grip is
  supported by the side of the second finger or the base of the thumb, and the
  shaft of the tool is 'external' to the hand (e.g. the way that a pen is held –
  see Figure 12.4).

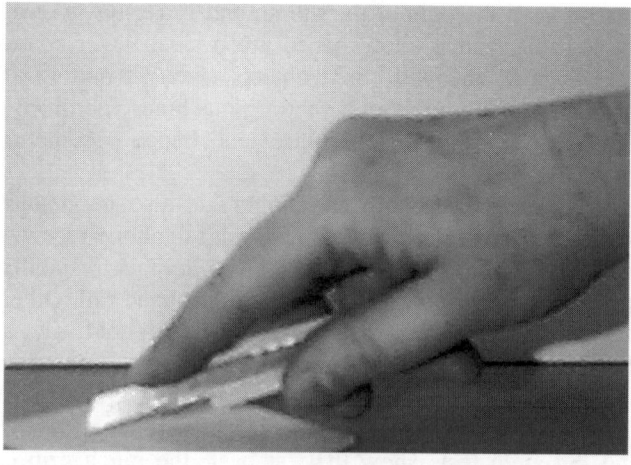

**Figure 12.3**  An internal precision grip is used to hold a knife to cut.

**Figure 12.4**   An external precision grip is used to hold a pen to write.

## 12.6.2   Power Grip (Cylindrical)

The power grip is the most powerful of all the hand grips because it allows the person to use the maximum force that can be developed by the hand. This grip is characterized by all of the fingers and the thumb flexing around an object. The power grip is a very common grip when we grasp a tool such as a power drill, a hammer or a wrench, and when we grab hold of objects, such as holding onto a rail. The power grip is often used when we carry objects, such as carrying a pipe or holding an object by its handle. It is also often used when we rotate objects, such as turning a screwdriver or a door knob. Most larger hand-tools require the use of a power grip to hold the tool or to hold onto a device (e.g. holding onto a tubular handle on say a lawn mower while this is being operated).

Although it is the most powerful of the grips, wrist orientation and grip span affect the force that can be generated by the hand. The greatest force can be exerted with a power grip when the hand is in a neutral position. As wrist orientation deviates from neutral so power grip force declines substantially. When the wrist is extended by 45° the grip strength falls to 75 per cent of maximum, and when it is flexed 45° grip strength falls to 60 per cent maximum. With the hand in maximum flexion (around 65°) grip strength is only 45 per cent of that for a straight hand. Likewise, the more hand goes into ulnar deviation the less grip force that can be generated, and at 45° ulnar deviation grip strength is only 70 per cent of maximum. The hand is less able to deviate radially (towards the thumb) and at 25° radial deviation grip strength is only 80 per cent of maximum. Thus, in designing a hand-tool that requires

**Table 12.3**  Optimum and maximum spans for a power grip for a 50/50 mixed workforce (cm)

|           | 5th %ile | 50th %ile | 95th %ile |
|-----------|----------|-----------|-----------|
| Optimum   | 4.5      | 5.5       | 5.9       |
| Maximum   | 9.5      | 11.0      | 13.0      |

*Source:* Eastman Kodak Co., 1983

forceful operation, such as a hammer or a wrench, keeping the hand as straight as possible is a significant factor for reducing fatigue and the risk of injury. Grip span also affects the strength of a power grip. If the span is too small or too large then the force that can be applied decreases. Optimum grip spans for a mixed workforce are given in Table 12.3.

### 12.6.3  Oblique Grip

This is a variant of the power grip that is characterised by the hand gripping across the surface of an object with the hand deviated from a neutral position. The oblique grip is used in many situations, such as when we use a wrench, spanner or screwdriver, pull on a rope or chain, and when we carry a tray with handles at either side, or grip a car steering wheel. Because the hand is ulnar deviated with an oblique grip, at its best the force generated is around 65 per cent of the grip strength of a neutral hand power grasp. As with other grips, the strength of this grip depends on the hand span required (Table 12.4).

### 12.6.4  Hook Grip

This is another variant of the power grip and it is characterised by a flat palm of the hand and curled fingers, such as the grip we use when we lift a box or carry

**Table 12.4**  Optimum and maximum spans for an oblique grip for a 50/50 mixed workforce (cm)

|           | 5th %ile | 50th %ile | 95th %ile |
|-----------|----------|-----------|-----------|
| Optimum   | 3.6      | 4.5       | 5.8       |
| Maximum   | 9.5      | 11.0      | 13.0      |

*Source:* Eastman Kodak Co., 1983

a bucket. In this grip the thumb is passive and used primarily to stabilise the load. The hook grip is most effective when the arms are down at the body side. At its optimum, the hook grip equals the strength of the power grip. Like the power grip, grip strength depends on hand span, and around 5 cm is the optimum diameter for an object being grasped. Grip strength decreases if the product has a very narrow handle, and, where possible, rigid handles should be avoided.

### 12.6.5  Palm-up/Palm-down Grips

These grips are commonly used when lifting or carrying cylindrical products, such as cans or jars. If we carry a product using the palm down grip then the stronger arm muscles of the biceps are not optimally positioned, and if the object weighs more than 0.5 kg, we will find it difficult to maintain this grip. The palm-up grip strength depends on the vertical position of the arms: maximum strength occurs with the arms around elbow level. Once the arm is positioned above elbow level, palm-up grip strength decreases.

Although five grips have been described separately, it is worth remembering that in many occupational situations the hands adopt different but complementary grips to perform tasks, such as holding a nail by the left hand with a pinch grip while holding a hammer with the right hand with a power grip. In the design of any hand-operated device, specifying the most appropriate grip for the device during its normal operation and also while the device is being carried is of paramount importance in determining the most appropriate form of the design for safe and efficient working.

## 12.7  Hand-Tools

Workers in many agricultural and industrial occupations make intensive use of hand-tools, and poor tool design can adversely affect job performance and lead to increased accidents and injuries. Analysis of injury data suggests that poor hand-tool design is a significant factor in around 10 per cent of all compensable and disabling injuries among US workers, and around 2 per cent of fatalities (Lewis and Narayan, 1993). Each year, injuries associated with poor hand-tool design cost the USA more than $10 billion. In addition to these direct injury concerns, poor hand-tool design is also implicated in a variety of insidious and progressive disorders, such as cumulative trauma disorders of the hands and wrist and vibration white finger. When ergonomically well designed hand-tools are correctly used they can reduce the risks of hand, wrist and forearm occupational injuries (Sperling et al., 1993). Such designs will also make tool use more comfortable and benefit job performance.

The ergonomic design of hand-tool handles requires careful consideration of a variety of factors (Lewis and Narayan, 1993). The most important physical factors to improving the ergonomic design and safety of the tool handle are:

- *Product shape*   This refers to both the overall dimensions of the product and the shape of the product area that will be gripped or operated by the hand. If the product has to be held then the grip area may be either an integral component of the product shape (e.g. a laser pointer, knife, hammer, cellular telephone) or a separate feature, such as a handle (e.g. on a coffee mug, briefcase, power drill). Product shape can inadvertently exclude some users, for example it is extremely difficult to use with the left hand scissors designed for right-handed operation. In many instances it is desirable to design the tool for operation with either hand. The overall shape of the product will affect aesthetic judgements and product desirability, often irrespective of its ergonomic design performance. Great products, however, are those that successfully integrate overall aesthetics with good ergonomics. Product shape also affects injuries and safety if the product presents other hazards, such as sharp corners or sharp edges.

- *Handle design*   For many tools, the handle is critical to successful operation. Good ergonomic design requires consideration of at least the following factors:
  - handle diameter
  - handle shape
  - handle length
  - handle texture
  - handle material
  - handle balance.

In cross-section, the size of the handle must fit the hands of those who may use the tool. Since a handle is meant to be grasped with a power grip, the handle diameter should be in the range of 3–5 cm. The handle should also be shaped to fit the contours of the palmar surface of the hand when the desired grip is used with the desired hand. This means that the handle should in some way be sculpted to the palm to the angle of the thenar eminence when the thumb is curled for a power grip. Handle length should be long enough that the handle passes completely across the palm, so that its end cannot cause compression to the palm. The handle should preferably be textured to improve grip and prevent slippage should it become damp with water, oil or sweat. A well designed grip cover also helps to spread the forces over the surface of the palm and absorb any vibrations or shocks to the tool. If the handle is made from metal or a material that readily conducts heat, then a grip cover can help to keep the hand cool when conditions are hot, and warm when conditions are cold. Sometimes hand-tools have an insulated grip cover to protect the user against electric shock. The handle can be designed so that it protects the hand with a guard or so that it attaches to the hand in some way to allow the user to let go of the tool without dropping it. Finally, the handle must be made from a material that is strong enough for the forces that will be generated by tool use.

In addition to this, other important considerations are the amount of force with which the tool is used, the weight of the tool, the degree of precision that is required when the tool is being used, the length of time that it is used, whether it is used by a bare or gloved hand, and what the prevailing climate conditions are, that is whether these are hot or cold, wet or dry, clean or dirty.

## 12.8  Ergonomic Principles for Good Design

All of the principles for good ergonomic design rest on five fundamental requirements that apply to any hand-operated device (adapted from Putz-Anderson, 1988). These requirements are that, to the extent that is practicable, the device should:

- Minimize high grip surface contact forces.
- Minimize static postural loading and pinch points when using the device.
- Minimize awkward joint positions and maximize neutral postures.
- Minimize repetitive forceful finger action with the hands in a deviated posture.
- Minimize the extent to which the device vibrates the hand.

Several specific design principles follow from these fundamental requirements.

### 12.8.1  Hand-Tool Design Principles

A fundamental requirement for all hand-tools is that a grip surface needs to be provided. This is usually in the form of a handle that isolates the hand from the working area of the tool. A well designed handle can enhance tool use by increasing grip stability and comfort. A well designed handle should achieve the following:

- The handle/tool should be shaped to avoid extremes of wrist deviation. This can be achieved by:
  - a bend in the tool handle;
  - using a pistol shape rather than a straight cylindrical design;
  - choosing an appropriate handle diameter or thickness. Grant *et al.* (1992) found that strength is maximized, and forearm muscle activity

and effort are minimized when the handle diameter is about 1 cm smaller than the inside grip diameter of the hand;

- padding the handle (see below).

■ The handle/tool should be shaped to assist the grip:

- padding the handle helps to reduce grip force if this results in a handle diameter between 2.5 and 7.5 cm. The padding should not be deeply contoured because this can create pressure points in the hands with forceful gripping. If the padding is slippery it will also reduce grip strength. Fellows and Frievalds (1991) compared the effects of foam rubber and wooden grip handles on six garden tools – leaf rake, garden rake, hoe, shovel, hedge shears and loppers. They found that users preferred the foam rubber handles, even though they reported less 'feeling' with these, and the deformation of the foam rubber handles required a greater grip force;

- a contoured grip can sometimes help to increase grip strength, such as the handle on a ski pole. At other times a contoured grip can interfere with the ability to grip the item, e.g. scissors often have a heavily contoured grip that may fit an adult right hand but which may cause difficulties if you have small or large hands or if you are left handed. The grip on a pair of scissors can be reshaped to provide universal operation (Figure 12.5).

■ Use universal designs that can accommodate ambidextrous operation and tool use by different user groups, such as children or those whose mobility is impaired.

■ Use designs that minimise shoulder abduction away from the body during normal tool use. Studies show that as shoulder abduction increases beyond 30° there is a progressive increase in the rate at which the shoulder muscles will fatigue. The tool should be shaped to allow normal operation with the arms kept down by, and as close to the body as possible.

■ Design to reduce hand-tool weight and to ensure that the centre of gravity passes through the grip. For example, a power drill with a short handle and a heavy body will require more strength to grip and operate than one with a longer handle and a smaller body.

There are many examples of better ergonomic designs for hand-tools that have resulted from the application of the principles described above. These examples include improved designs for a knife, pliers, soldering iron, tennis racquet, pizza cutter, ice cream scoop and shovel, among others. Konz (1990) provides a good overview of several of these designs, and Bobjer (1989) describes the benefits of redesigning two types of knife for meat cutters. To illustrate why ergonomic design improvements are important, let us briefly consider studies on alternative hammer designs.

(a)

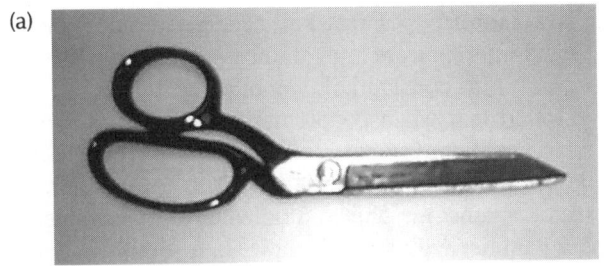

(b)

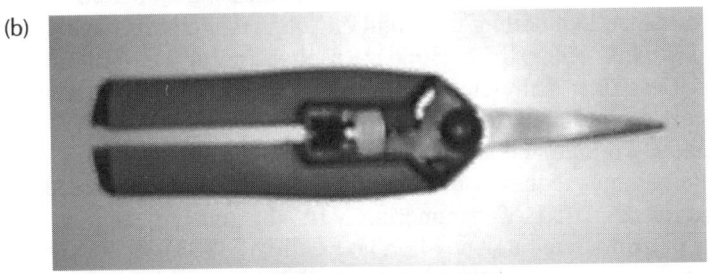

(c)

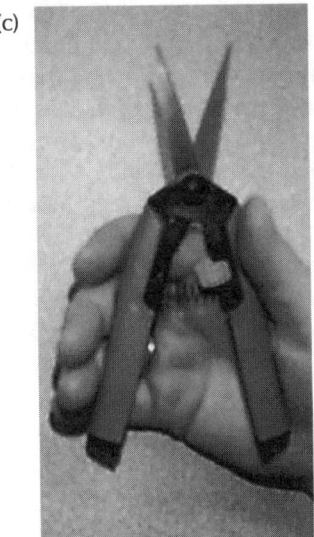

**Figure 12.5** The grip on a pair of scissors can be reshaped to provide universal and comfortable operation, left- or right-handed. (a) Conventional design for a pair of scissors. (b) Ergonomic redesign for universal operation. (c) Ergonomic redesign also improves hand grip.

The hammer is a familiar tool that has been used since ancient times. In medieval times, hand injuries were known to be associated with stone masonry, a profession requiring a considerable amount of hammer use. Recently, researchers have found that when people use a straight-handled hammer they make repetitive, forceful movements with their wrist bent alternately from extreme radial deviation as the hammer is raised, to extreme ulnar deviation as the hammer is struck against the object. Studies have investigated the potential benefits of using a hammer with a bent handle. In a series of studies, Konz and his colleagues have investigated the effects of various bent hammer designs on work performance in a nailing task, and on subjective comfort and preferences (Konz and Streets, 1984; Konz, 1986). Results suggest that, at least for short-term work (less than 5 minutes), college students prefer a bent-handle (10°) design to a straight-handle design, even though there were no differences in nailing accuracy or nail depth between designs. Other studies have shown that a bent-handle hammer requires substantially less ulnar deviation to operate, and at the end of a period of hammering with this design there is less of a decrement in grip strength (Knowlton and Gilbert, 1983). Schoenmarklin and Marras (1989a,b) compared the use of a straight-handled hammer with that of two different bent-handle hammers (20° or 40° bend) on surfaces at different orientations – vertical, to simulate a wall, and horizontal, to simulate a bench or floor. No performance differences were found between the hammer designs; however, the total amount of ulnar deviation was less with the bent-handle designs. When hammering a nail into a wall (vertical), people generated less driving force (no benefits of gravity) and made more misses than when hammering horizontally, but performance was best with the 20° bent-handle hammer. No evidence was found that any of the designs required less muscular effort to perform the same amount of work. Thus, results from these studies suggest that a hammer with a handle bent between 10° and 20° will be preferred and will help to reduce the postural risks of a hand wrist injury. Although there are benefits to a bent-handle hammer design, such designs are still not in widespread use.

### 12.8.2  Tiltdown Keyboard Holder

Computer keyboards are commonplace in most companies in developed countries. First developed around 1868 by Charles Latham Sholes for the typewriter that he invented, the QWERTY key layout for the alphabetic keys has become standard in many countries. Although other keyboard layouts, such as the Dvorak layout, have been touted as superior, careful examination of the limited research suggests that the story of QWERTY's inferiority is a myth (Liebowitz and Margolis, 1996). Compared with modern keyboards, the mechanical typewriter had relatively flat keys that required greater force to depress and were in a stepped design. Typists were trained to type while keeping their wrists in a straight, neutral posture. With modern electronic keyboards, the keys are usually in a sloped arrangement and less force is required to activate a key. The physical design of the modern computer keyboard as a product is important to performance and for cumulative trauma injury risks.

There is growing concern worldwide that intensive keyboard use may increase the risk of developing a cumulative trauma injury, such as carpal tunnel syndrome (CTS). Occupational CTS appears to be caused by cumulative damage to the finger tendons and/or the median nerve as these structures pass through a 2–3 cm long, narrow, rigid channel in the wrist – the carpal tunnel. With the hand oriented palm down, the roof of the carpal tunnel is formed by the arch of the carpal bones and the floor by the tough transverse carpal ligament. The carpal tunnel contains the tendons for the fingers, the radial artery, and the median nerve which transmits sensation for the thumb and the first 2.5 fingers. Sensation for the remaining 1.5 digits is transmitted via the ulnar nerve which runs outside of the carpal tunnel.

As the hand deviates from the normal range either horizontally towards the thumb (radial deviation) or towards the little finger (ulnar deviation), or vertically up (extension) or down (flexion), the pressure in the carpal tunnel increases (Armstrong et al., 1984), and accelerations from extension to flexion are thought to pose the greatest risk for CTS (Marras and Schoenmarklin, 1991). When the hand is in a wrist neutral position (i.e. no vertical or horizontal deviation) there is the minimum pressure on the tendons and the median nerve in the carpal tunnel. With occupational overuse of the fingers, minor trauma to the tendons and the sheaths may accumulate and eventually produce CTS. Repetitive movements with the hands in a deviated posture accelerate the onset of CTS (Chaffin and Anderson, 1984; Putz-Anderson, 1988). As the tendons or their sheaths become irritated and inflamed, the resulting swelling increases the pressure on the median nerve, which initially causes tingling, then numbness, and eventually disabling pain when the fingers are moved. Computer users are particularly at risk because of the large number of movements which the fingers may make in a short time (e.g. a data entry worker who averages 13 000 keystrokes per hour will make over 0.5 million finger movements per week). In short, the major occupational risk factors for CTS are poor hand/wrist posture causing an increased pressure in the carpal tunnel, keying force and insufficient pauses to allow time for tissue recovery and trauma repair.

The use of a conventional keyboard can cause ulnar deviation of both hands, especially the right because the alphanumeric part of a 101 key keyboard is off-centre left of the midline and workers tend to centre themselves on the whole keyboard rather than the alphanumeric part. The positive keyboard angle places the hands in an extended posture. Conventional computer keyboards usually are designed to slope at a positive angle of around 15° to the horizontal. When placed on a desktop or conventional flat, height-adjustable keyboard tray, extremes of wrist extension beyond 20° can readily be observed for many users during typing (Hedge and Powers, 1995; Hedge et al., 1996). Such extremes

**Figure 12.6**   Computer worker using a downward-tilting keyboard platform for improved wrist posture while typing.

can also be observed when users raise their fingers and hands off the keyboard during the frequent microbreaks, which characteristically occur between bursts of typing. Together, these postural deviations and the repetitive nature of typing increase the risks of occupational CTS. Reducing the keyboard angle using a negative slope keyboard holder, called a tiltdown keyboard system, and lowering the height of the keyboard relative to the normal worksurface, allows a geometric design solution which theoretically minimizes hand excursions beyond the neutral zone for the wrist during typing (Figure 12.6). Studies have confirmed the efficacy of the tiltdown keyboard arrangement for improving posture in both laboratory (Hedge and Powers, 1995) and field research (Rudakewych *et al.*, 1994; Hedge *et al.*, 1996).

## 12.9  Conclusions

This chapter has summarised a series of important ergonomic considerations for the design of hand-operated products, focusing on a consideration of how the design of the product fits the capabilities of the hands so that comfort, ease and safety of product use are maximised and performance is enhanced. The criteria that have been presented can be used to evaluate the degree to which the design of existing products can be judged to be ergonomic. These criteria should also be used to assist in the development of new products to ensure that the requirements for an ergonomic product are met.

## References

ARMSTRONG, T. J., CASTELLI, W. A., EVANS, F. G. and DIAZ-PEREZ, R. (1984) Some histological changes in carpal tunnel contents and their biomechanical implications, *Journal of Occupational Medicine*, **26** (3), 197–201.

BOBJER, O. (1989) Ergonomic knives. In Mital, A. (ed.), *Advances in Industrial Ergonomics and Safety*. London: Taylor & Francis, pp. 291–8.

CHAFFIN, D. B. and ANDERSON, G. (1984) *Occupational Biomechanics*. New York: John Wiley.

DAHLIN, L., NORDBORG, C. and LUNDBORG, G. (1991) *Journal of Hand Surgery*, **16A**, 753–8.

EASTMAN KODAK COMPANY (1983) *Ergonomic Design for People at Work*, vol. 1, Belmont, Ca.: Lifetime Learning Publications, pp. 140–53.

(1986) *Ergonomic Design for People at Work*, vol. 2, New York: Van Nostrand Reinhold, pp. 348–59.

FELLOWS, G. and FREIVALDS, A. (1991) Ergonomics evaluation of a foam rubber grip for tool handles, *Applied Ergonomics*, **22** (4), 225–30.

GELBERMAN, R. H., SZABO, R. M. and MORTENSON, W. W. (1984) Carpal tunnel pressures and wrist position in patients with Colle's fractures, *Journal of Trauma*, **24** (8), 747–9.

GRANT, K. A., HABES, D. and STEWARD, L. L. (1992) An analysis of handle designs for reducing manual effort: the influence of grip diameter, *International Journal of Industrial Ergonomics*, **10** (3), 199–206.

HARGENS, A. R., ROMINE, J. S., SIPE, J. C., EVANS, K. L., MUBARAK, S. J. and AKESON, W. H. (1979). Peripheral nerve-conduction block by high muscle-compartment pressure, *Journal of Bone and Joint Surgery*, **61A** (2), 192–200.

HEDGE, A. and POWERS, J. R. (1995) Wrist posture while keyboarding: effects of a negative slope keyboard support system and full motion forearm supports, *Ergonomics*, **38**, 508–17.

HEDGE, A., McCROBIE, D., MORIMOTO, S., RODRIGUEZ, S. and LAND, B. (1996) Painfree computing: use of a preset tiltdown keyboard system and new tools for visualizing wrist postures lead the fight against carpal tunnel syndrome. *Ergonomics in Design*, **4** (1), 4–10.

KAPANDJI, I. A. (1982) *The Physiology of the Joints*, vol. 1. New York: Churchill Livingstone.

KNOWLTON, R. G. and GILBERT, J. C. (1983) Ulnar deviation and short-term strength reduction as affected by a curve-handled ripping hammer and a conventional claw hammer, *Ergonomics*, **26**, 173–9.

KONZ, S. (1986) Bent hammer handles, *Human Factors*, **28**, 317–23.

KONZ, S. (1990) *Work Design: Industrial ergonomics* (3rd edn). Worthington, Ohio: Publishing Horizons, pp. 237–58.

KONZ, S. and STREETS, B. (1984) Bent hammer handles: performance and preference, *Proceedings of the Human Factors Society 28th Annual Meeting*, vol. 1. Santa Monica, Calif.: Human Factors and Ergonomics Society, pp. 438–40.

LEWIS, W. G. and NARAYAN, C. V. (1993) Design and sizing of ergonomic handles for hand tools, *Applied Ergonomics*, **24** (5), 351–6.

LIEBOWITZ, S. and MARGOLIS, S. E. (1996) Typing errors, *Reason*, June. Also available from Reason Online at http://www.reasonmag.com/9606/Fe.QWERTY.html.

MARRAS, W. S. and SCHOENMARKLIN, R. W. (1991) Wrist motions and CTD risk in industrial and service environments. In: Quéinnec, Y., Daniellou, F. *et al.* (eds), *Designing for Everyone: Proceedings of the Eleventh Congress of the International Ergonomics Association, Paris, 1991*, vol. 1. New York: Taylor & Francis, pp. 36–8.

PUTZ-ANDERSON, V. (1988) *Cumulative Trauma Disorders: A Manual for Musculoskeletal Disorders of the Upper Limbs*, New York: Taylor & Francis.

REMPEL, D. and HORIE, S. (1994) Effect of wrist posture during typing on carpal tunnel pressure. In: Grieco, A., Molteni, G., Occhipinti, E. and Piccoli, B. (eds), *Work with Display Units '94, Proceedings of the Fourth International Scientific Conference, University of Milan, Italy*, vol. 3, C27–C28.

REMPEL, D., HORIE, S. and TAL, R. (1994) Carpal tunnel pressure changes during keying. In *Proceedings of the Marconi Keyboard Research Conference*, UC San Francisco, Ergonomics Laboratory, Berkeley, 1–3.

ROBBINS, H. (1963) Anatomical study of the median nerve in the carpal tunnel and etiologies of the carpal-tunnel syndrome, *Journal of Bone and Joint Surgery*, **45A** (5), 953–66.

RUDAKEWYCH, M., VALENT, L. and HEDGE, A. (1994) Field evaluation of a negative slope keyboard system designed to minimize postural risks to computer workers. In Grieco, A., Molteni, G., Occhipinti, E. and Piccoli, B. (eds), *Work with Display Units '94, University of Milan, Italy*, vol. 3, C17–C19.

SCHOENMARKLIN, R. W. and MARRAS, W. S. (1989a) Effects of hand angle and work orientation on hammering. I: Wrist motion and hammering performance, *Human Factors*, **31**, 397–411.

SCHOENMARKLIN, R. W. and MARRAS, W. S. (1989b) Effects of hand angle and work orientation on hammering. I: Muscle fatigue and subjective ratings of body discomfort, *Human Factors*, **31**, 413–20.

SPERLING, L., DAHLMAN, S., WIKSTRÖM, L., KILBOM, A. and KADEFORS, R. (1993) A cube model for the classification of work with hand tools and the formulation of functional requirements. *Applied Ergonomics*, **24** (3), 212–20.

# CHAPTER THIRTEEN

# Development of comprehensible warning symbols for use on child-care products

MONICA TROMMELEN[1] and HARM J. ZWAGA[2]

[1]*Faculty of Social Sciences, Leiden University, The Netherlands*
[2]*Faculty of Social Sciences, Utrecht University, The Netherlands*

## 13.1 Introduction

### 13.1.1 Background

The research discussed in this chapter has been initiated to support the work of Project Group 5 (PG5) of the European Committee for standardization (CEN/TC252/WG6/PG5) and is fully reported by Akerboom *et al.* (1995). The aim of PG5 is the harmonisation in Europe of product information for child-care products. Studies previously conducted to support the work of PG5 were aimed at product information in general and at the development and evaluation of comprehensible warning sentences (Hagenaar and Trommelen, 1992; Trommelen, 1994, 1995). Here, we discuss the work for PG5 on the systematic development and testing of warning symbols for child-care products.

### 13.1.2 Warnings

Effective warnings should result in safe behaviour, leading to a reduction in the number of accidents. Unfortunately the response rate to warnings is usually low. Research shows that many either do not notice warnings, fail to read them, or do not comply with them (e.g. Dorris and Purswell, 1978; Friedmann, 1988, Otsubo, 1988). The question is how to raise the impact of warnings.

Edworthy and Adams (1996, p. 3) argue that a warning (sign) should be thought of as an artefact that represents the risk associated with the hazardous situation. In order to do so, a warning usually serves as an alerting function and an informing function. The alerting aspects of a warning serve as an indication of a hazard and the severity of a hazard. Signal words, colours, symbols and sound are examples of alerting elements in a warning. When used effectively, these elements require little conscious information processing: they are almost spontaneously understood. The informing aspects of a warning give indications on how to handle a hazardous product or how to act in a hazardous situation. The two aspects of a warning are often difficult to separate. For example, both alerting cues and actual information can be embodied in the phrase 'Handle with care', provided a proper layout, type size, etc. are chosen.

Perceived hazardousness and perceived costs of compliance are the most influential factors determining the consumers' motivation to pay attention to a warning and to respond to it in an appropriate way (Dingus *et al.*, 1991; Wogalter *et al.*, 1991). To affect the beliefs concerning the hazardousness of a product, the consumer should be adequately informed about the existence, nature and magnitude of the hazard(s). Moreover, information on the severity of the consequences should be provided (e.g. Vaubel and Brelsford, 1991; Trommelen and Akerboom, 1997) to make people understand why they should perform or avoid certain actions.

Thus, to communicate safety information effectively, a warning should comprise the following (e.g. Sanders and McCormick, 1993, p. 683; Wogalter *et al.*, 1987):

- A *signal word* to convey the gravity of the risk.
- An indication of the *hazard*.
- The possible *consequences* in terms of injuries.
- *Instructions* as to how to avoid injuries.

Research shows that if one or more of these four kinds of information is, in addition, presented as a graphic symbol, compliance will increase (e.g. Jaynes and Boles, 1990; Young and Wogalter, 1990). Edworthy and Adams (1996, Chapter 3) in their detailed overview of the use of symbols in warnings, stress the point that iconic information in the form of a warning symbol can at least have an alerting function even if the consumer does not understand the meaning of the symbol (no informing function).

### 13.1.3  Graphic Symbols

The use of symbols has some important advantages. Firstly, symbols are not subject to the limitations of written text: there is no need to know the language to understand the meaning of the symbol. Secondly, symbols need less space than text to be legible from the same distance. Thirdly, symbols usually attract the attention more easily than printed text, because of their shape, size or colour. Fourthly, symbols with an expressive image design can have a much higher impact than printed text, i.e. the information is more quickly and better processed.

A disadvantage of symbols, or rather the use of symbols, is that often symbols are used which are difficult to understand. There clearly is a need to verify the comprehensibility of public information symbols.

Symbols can be descriptive, prescriptive or proscriptive (Boersema and Zwaga, 1989). Descriptive warning symbols indicate the existence of a hazard (e.g. fire, poison); prescriptive symbols specify a positive course of action to be taken in association with the hazard (e.g. wear protective gloves); proscriptive symbols indicate which behaviour should not occur (e.g. do not drink). Descriptive symbols are usually best understood (Easterby and Hakiel, 1981). If a warning should have to consist of symbols only, a useful solution is probably to combine different symbols into one warning message, e.g. a symbol for the hazard adjacent to a symbol instructing what to do. The feasibility of this approach has been demonstrated by Zwaga *et al.* (1991) when developing and evaluating a set of warning symbols.

### 13.1.4 Aim of the Study

The selection of comprehensible symbols to be implemented in five warnings is used to demonstrate the feasibility of a procedure to develop graphic symbols with an objectively verified level of comprehensibility. The selection of the warnings is based on an analysis by Trommelen (1994) of 15 European (draft) standards for the safety of child-care products to identify the main hazards of these products.

Warnings symbols should be developed for the following referents.

- Keep this (plastic) cover away from your child to avoid suffocation.
- Do not use this (product name) once your child is older than *x* years of age; taller than *x* cm; weights more than *x* kg. The (product name) may collapse/fall over. Your child may be hurt.
- Never leave your child unattended in/with this (product name). Your child may be hurt.

Subsequently the following abbreviations will be used for these referents: suffocation; age/height/weight; unattended.

## 13.2 General Description of and Rationale for the Test Procedure

For the development of the warning symbols, an iterative test procedure is used. The procedure adheres to the main requirements of the procedure prescribed in the standard ISO/DIS 9186 Rev. 1995-01-03 *Procedure for the development and testing of public information symbols* (ISO, 1995). The aim of the test procedure is to select symbols with a verified level of comprehensibility in an efficient and objective way. The test procedure consists of three phases:

- Selecting symbol variants for testing.
- Testing of selected symbol variants in estimation test(s).
- Testing the best symbol variants in comprehension test(s).

With the test(s) in the third phase, the level of comprehensibility of a symbol is determined objectively. The results of the comprehension test are used to decide on the acceptance or rejection of proposed symbols. The comprehension test has one major disadvantage: all proposed symbols for a single referent have to be tested in separate tests. To save time and money, an objective way has to be found to reject in advance the less promising symbol variants. The estimation test has been developed for this purpose. Its applicability and effectiveness is described in detail by Brugger (1990). The aim of the test is to screen the proposed symbols on comprehensibility. Because of this screening no stringent limits have to be put on the number of variants proposed for testing.

The three phases of the procedure are described below.

### 13.2.1    Phase One: Selecting Symbol Variants for Testing

The aim of the first phase is to collect for each of the referents as many symbol designs as possible. If not enough symbol designs can be located, a production test can be used to assist in generating useful symbol variants. In a production test, members of the target population are asked to design symbols for the different warnings. The resulting set of drawings can be used to develop symbol variants. Either the ideas expressed in a drawing or specific figurative elements from the drawings can be used to develop symbol variants.

### 13.2.2    Phase Two: Testing Symbol Variants in Estimation Tests

The aim of the second phase of the test procedure is to select the most promising symbols from the total set of symbol designs collected in the first phase.

In an estimation test, the respondents are asked to estimate the percentage of the adult population of their native country that they think will understand the meaning of the different symbol variants for each of the referents. The median of the estimates of the respondents for a symbol is its estimated comprehension score. According to the ISO standard it can be assumed that symbols with an estimation score above 85 per cent will pass the comprehension test and can be accepted without further testing. Symbols with an estimation score below 45 per cent will fail the comprehension test and can be rejected. The level of comprehension of symbols with an intermediate estimation score (between 45 and 85 per cent) will have to be tested in a comprehension test to determine if they meet the acceptance criterion.

This study differs from the ISO standard because more stringent requirements are needed for warning symbols than for public information symbols. Therefore, it was decided that symbols with an estimation score above 85 per cent on comprehension would be tested as well.

For a symbol to be selected as a potentially good warning symbol it should in fact meet two conditions. The median of the distribution of the estimates should be high (over 90 per cent) and, in addition, the shape of the distribution of the estimates should be positively skewed. This means that a large majority of the estimates are very high, some have a medium value and there are only a few very low estimates. The other way around, a potentially bad symbol has a very low median in combination with a highly skewed distribution, but now the distribution is located at the low side of the range. This approach provides useful guidance when analysing the results of an estimation test. It should be realised, however, that in practice other factors besides median and shape of the distribution play a role in the definite selection of symbols for further testing. Factors such as the simplicity of a symbol or coherence of the resulting symbol set can affect choices. This means that, although a symbol with a high median and a nicely skewed distribution is available, a symbol with a lower median and a less skewed distribution is selected, or also selected, because it is less detailed and easier to design in the same style as the other potentially acceptable symbols selected for further testing. It is justifiable, at this stage, to diverge from the adopted rules because the final decision to accept symbols lies with the comprehension test.

If, for all referents, one or more symbols meeting the criterion are available, the second phase of the procedure is completed. Otherwise the estimation test has to be repeated with new or adapted symbols until, for all referents, potentially good symbols are available. In the study reported here, two successive estimation tests were needed (see Section 13.4).

### 13.2.3 Phase 3: Testing Symbol Variants in Comprehension Tests

In this last phase of the procedure the variants selected with the estimation tests are further tested to verify their level of comprehensibility. In the comprehensibility test, the respondents are first explicitly told about the context of use of the symbols, then they are shown one symbol variant per referent and asked to write down what they think each symbol means. The percentage of correct interpretations of a symbol determines its comprehension score. For this study,

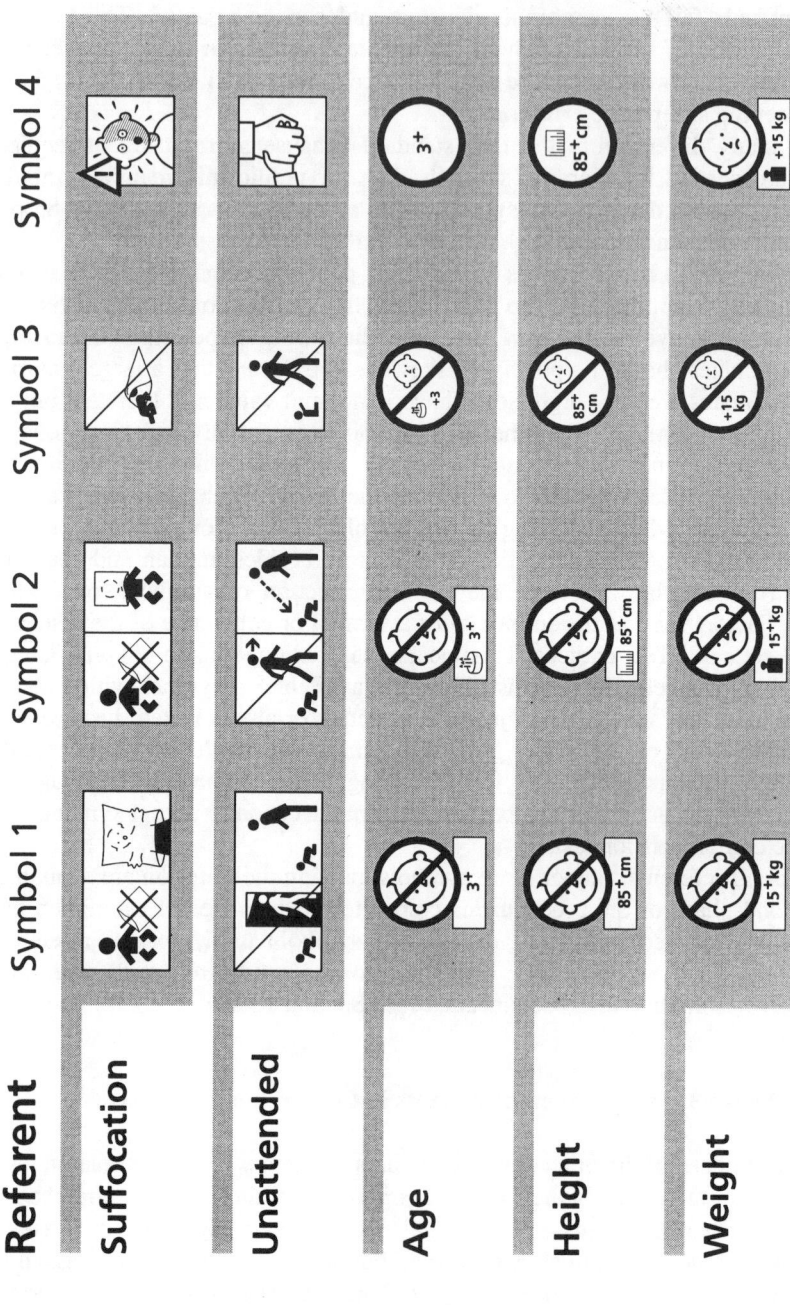

**Figure 13.1** Four examples of symbol variants (of a total of 55) for each of the five referents used in the first estimation test. Per referent, the number of variants varied from 10 to 13 symbols.

the acceptance criterion was set at 80 per cent correct interpretations with less than 4 per cent opposite interpretations. However, there is no agreed criterion for the acceptance of warning symbols. For public information symbols in general, the ISO standard prescribes an acceptance criterion of 66 per cent. For a detailed description of the procedure of the test see ISO (1995).

The results of a comprehension test, i.e. the interpretations given by the respondents, can also indicate why symbols are misunderstood and this information can then be used to adapt variants to improve their comprehensibility. Adapted symbols have to be tested in an estimation test, if there are many, or another comprehension test, to verify again their level of comprehension. This process has to be repeated until symbols with the required comprehension level are available for all referents. In this study, one set of three comprehension tests has been conducted in parallel (see section 13.5).

## 13.3  Initial Symbol Variants

Only a few symbol designs for the referents appeared to exist, so a production test was used to generate ideas for additional symbols. The results suggested several variants. Two different types of variant were designed: single symbols with only the do's and/or don'ts, or the hazards and paired symbols which combine either the do's and don'ts or the don'ts and the hazards. For 'suffocation', descriptive as well as proscriptive symbols were designed with, in some variants, the two combined in paired symbols. It was difficult to design descriptive symbols for 'unattended', because the hazards that this warning refers to are product-dependent. For a descriptive symbol this would mean that the image of the symbol should show the product. To avoid this problem, proscriptive and prescriptive symbols were designed, which were also combined in paired symbols.

It was difficult to design descriptive symbols for 'age/weight/height' because here, too, the image of a symbol should show the product it is used on. Therefore proscriptive and prescriptive symbols were designed for this warning as well.

Eventually 13 variants were chosen for 'suffocation' (11 single and 2 paired symbols), 12 variants for 'unattended' (6 single and 6 paired symbols), and 10 variants for 'age weight/height' (all single symbols). A selection of variants for the five referents are shown in Figure 13.1.

## 13.4  Estimation Tests

### 13.4.1  Procedure

In an estimation test the respondents are shown all the symbol variants for a referent. The symbols for a referent are depicted on a single page of the test form. A description of the referent and the context of use are printed in the middle of the page. The task of the respondents is described as follows: 'Could you give an estimate of the comprehensibility of each symbol on a page. Please do this by specifying the percentage of the (Dutch) adult population you think will understand the meaning of each of the symbols.' An example using symbol variants for another, unrelated, referent is given to explain the procedure.

Two estimation tests have been conducted. In the first test the symbol variants described in Section 13.3 were used. In the second estimation test, adapted and new symbols as well as symbols from the first test were used.

### 13.4.2   Results and Conclusions of the First Estimation Test

Data have been collected from two groups of respondents: 96 parents-to-be (average age 30 years) and 48 students (average age 23 years). The estimation scores of the groups did not differ significantly, although the scores of the students tended to be higher. The scores of the two groups have therefore been combined.

For age/weight/height similar variants came out best for each of the warnings (symbols 1 and 2 in Figure 13.1). Considering their median and the distribution of the estimates, two of the ten symbols for each of these three warnings were acceptable. The results were neutral (median 50–60 per cent and a rather flat distribution) for five of the symbols for age and weight, and for four of the symbols for height. The other of the ten symbols per referent had very low medians and negatively skewed distributions.

From the thirteen variants for suffocation only one (symbol 1 in Figure 13.1) had a positively skewed distribution and a high median (80 per cent). Two variants had negatively skewed distributions and the results of the other ten variants were neutral.

Three of the twelve variants for unattended had high medians and positively skewed distributions. The two best variants are shown as symbols 1 and 2 in Figure 13.1. Five of the variants had neutral results and the remaining four had negatively skewed distributions.

Summarising the results, it can be concluded that symbols for further testing are available for all referents. The results for suffocation and unattended also clearly show that paired symbols obtain better results than single symbols.

Even so, it was decided for a number of reasons to conduct a second estimation test. The main reasons were methodological. First, it was observed that whilst conducting the test many respondents seemed to react only to the general shape of the symbol image and to ignore what they called minor details. These were often the elements in the symbols that had been varied systematically to measure their effect on the estimation score. Most of the age/weight/height symbols had been devised in that way. This tendency to ignore details appeared to depend also on the number of symbol variants for a referent presented on a testsheet. Secondly, the fact that paired symbols came out best might be caused by the mixing of single and paired symbols on a testsheet. If paired symbols are better than single ones, the contrast between the two, shown on one page, might have instigated the respondents to stress the difference in comprehensibility between the two kinds of symbol. Finally, the results of the first test suggested that the comprehensibility of a number of symbols might be improved by making changes to their image content, or by making new combinations of symbols for paired ones. Several new designs have even been added because promising ideas for the image content of symbols presented themselves when discussing the results of the first estimation test.

### 13.4.3   Second Estimation Test

Considering the two possible sources of error mentioned, the following changes were made to the test procedure.

- To avoid presenting too many variants, the maximum number of alternative symbols on a page is six.
- If there are single and paired symbols for a referent, each kind is presented on a separate page. This to avoid the effects of the contrast between the two kinds of symbol on the estimates.

Another difference between the first and the second estimation test was that in the second test only the variants for age and not those for weight and height were tested. This was because the results of the first test had shown no marked differences between the estimates for the symbols of the three referents. The symbol variants for age have been chosen because they are the most critical ones. These symbols might be confused with the existing toy symbol. This symbol designates an opposite age limit (and the written warning 'Not suitable for children under $x$ years of age').

The sets of variants used in the second test included for each referent the symbol with the best results in the first estimation test. There were six symbols for age; three of them were adapted old symbols and two were new designs.

Five single and four paired symbols have been tested for the referent suffocation. Among them was one new single symbol and two new paired symbols. Both new paired symbols consisted of single symbols already used in the first estimation test.

For the referent unattended, four single and four paired symbols have been tested. One new single symbol was added. The other three symbols were slight modifications of symbols used in the first test. The four paired symbols all consisted of combinations of single symbols used in the first test.

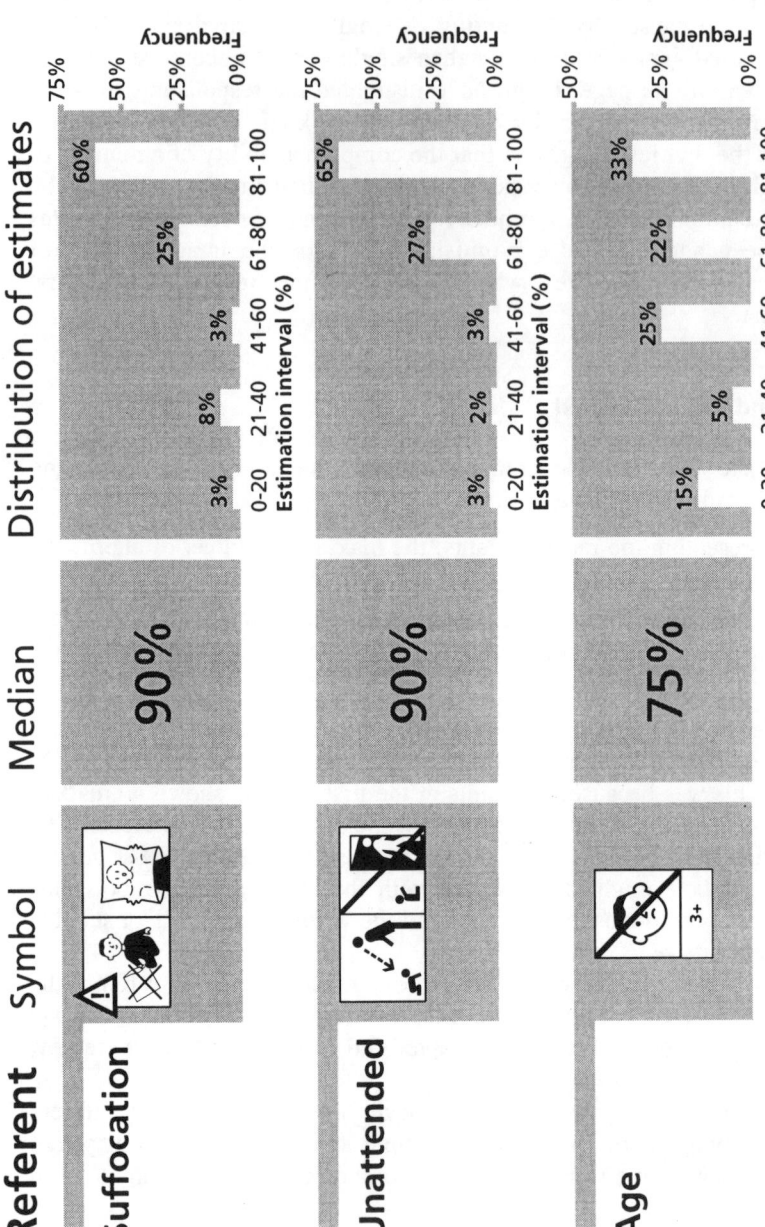

**Figure 13.2** Results of the second estimation test. For each of the three symbols selected for further testing in the comprehension test, two kinds of test data are shown: the median of the estimates and the distribution of the estimates. The symbols selected for suffocation and unattended obtained both a high median and almost ideally skewed distribution of the estimates. The results for age were less clear.

### 13.4.4  Results and Conclusions of the Second Estimation Test

The second estimation test was carried out with 120 Dutch respondents (students, 76 women and 44 men, average age 23 years).

Based on the results of the test, the three symbols shown in Figure 13.2 were selected as the best for further testing in the last phase of the procedure, the comprehension test. The selection of the best symbol for suffocation and unattended was straightforward; both symbols shown in Figure 13.2 had the highest median of the variants tested and an almost ideally skewed distribution of the estimates.

The results of the age symbols are not as clear as those of the other two referents. The symbol shown was chosen because it had the best combination of median value (75 per cent) and distribution of the estimates. Its distribution had the lowest number of estimates at the lower end of the range and the highest number at the highest end of the range.

## 13.5  Comprehension Tests

### 13.5.1  Procedure

As was mentioned in Section 13.2, the comprehension test is the last and most crucial step in the procedure to select comprehensible graphic symbols.

### Complicating Factors

There were two factors identified as necessitating a more complex method of investigation than usual. One was the fact that the symbols for weight and height had to be tested as well, and not only the symbol for age, as was the case in the second estimation test. The other factor was that the symbols eventually selected would be used as symbols for warnings on child-care products.

### The Symbols

Because the symbols for weight and height were built as equivalents to the age symbol, they look very similar. The upper part of these symbols is identical to the upper part of the age symbol (Figure 13.2) and the lower part, the rectangular box, says '15$^+$kg' and '85$^+$cm' respectively. In order not to bias the results, because of the resemblance of the three symbols, three different samples were used to collect independent data for these symbols.

### Additional Samples.

The second factor, i.e. the point that the symbols refer to warnings on child-care products, dictated that a prudent approach should be used with regard to the verification of the usability of the symbols. Therefore data have been collected not only from samples of the general public but also from three additional samples. These three samples consisted of members of a prominent part of the target group, i.e. parents-to-be.

## Test Procedure

The instruction for respondents can be summarised as follows. In a short introduction in which the field of application of the symbols is explicitly mentioned and the task is explained, the symbols are shown to the respondents. An example is given of the use of each of the symbols, mentioning that the symbol could be shown on a product itself or on the packaging. Next, respondents are asked to write down what they think is the meaning of each of the three symbols.

## Data Analysis

In the analysis of the results of a comprehension test, the decision whether a response is correct or not should be based on strict rules which are related to the purpose of the symbols. For the warning symbols of this test, a response is correct if two aspects are mentioned: a description of possible danger and a description of measures to avoid possible danger.

## 13.5.1   Results and Conclusions of the Comprehension Tests

### Samples

Data were collected from six independent samples. Three samples consisted of members from the general public and the respondents in the three other samples were parents-to-be.

Data for the symbols suffocation, unattended, and age were collected from a sample of 130 respondents from the general public (59 female, 71 male, average age 37 years) and from a sample of 25 parents-to-be (15 female, 10 male, average age 31 years).

Data for the symbols weight and height were collected from samples of 72 and 71 members of the general public respectively (in total 70 female, 73 male, average age 40 years) and from two samples of parents-to-be of 30 and 28 respondents respectively (in total 33 female, 25 male, average age 32 years).

### Results for 'Suffocation' and 'Unattended'

The results found for the two samples were similar with regard to correct, wrong and opposite responses. The symbols have a comprehension score of 92

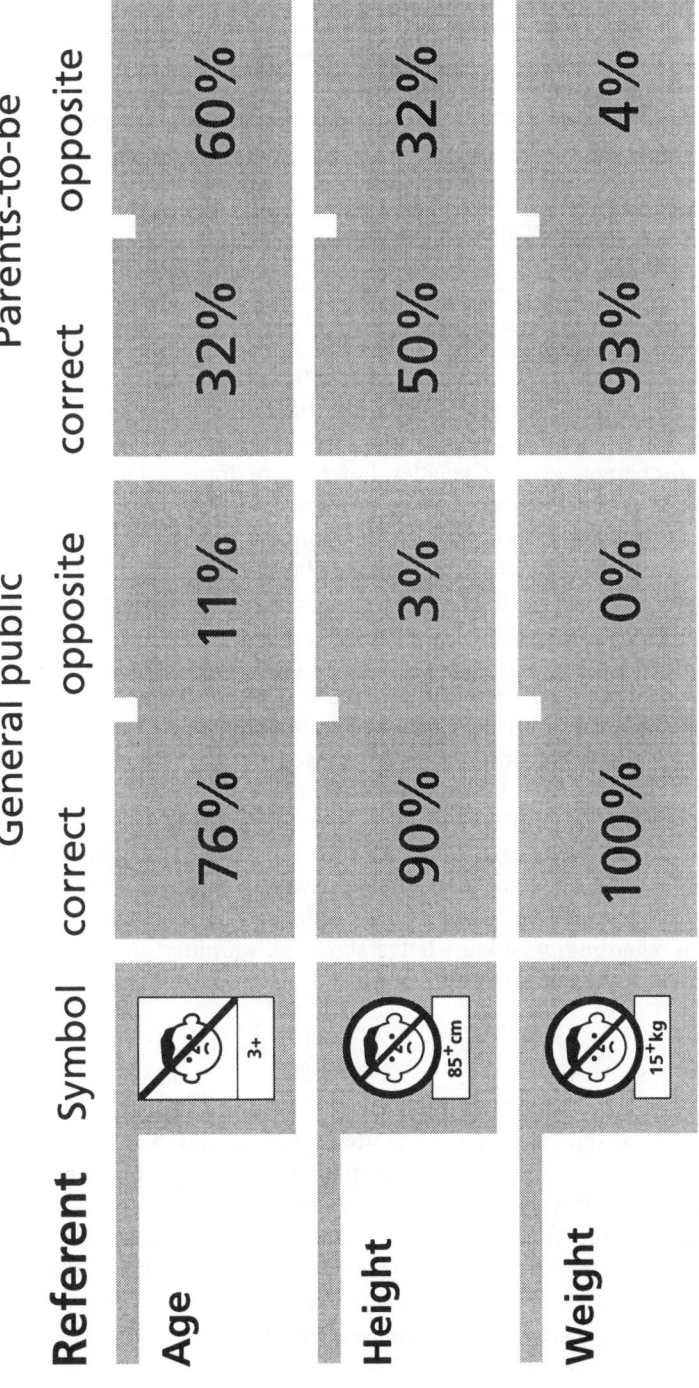

**Figure 13.3** Results of the comprehension test for the referents age, height and weight. Shown are the percentages correct and opposite meanings given by respondents from the general public and by samples of parents-to-be.

per cent and 84 per cent respectively. Neither triggers more than a negligible percentage of opposite responses.

Both symbols are sufficiently well understood to accept them as properly representing the warning they stand for. They meet the preset criteria chosen for this study (Section 13.2.3) and certainly meet the ISO acceptance criterion for public information symbols.

### Results for 'Age', 'Weight' and 'Height'

There are large differences between the comprehension scores for the three symbols. Considering only the percentages correct responses and percentages opposite responses (Figure 13.3), only the symbol for weight is acceptable. Inspection of Figure 13.3 indicates – and statistical analysis of the data confirms this – that the symbol for age is the most difficult to understand, followed by height and then weight. This pattern is the same for respondents from the general public and for parents-to-be. The latter, however, perform at a lower level.

These differences between results for the three symbols are difficult to interpret, especially as the symbols are very similar: one would expect the results to be similar too. The results of the age symbol may have been affected by the confusion with the toy symbol, which gives exactly the opposite message ('not suitable for children under three years of age'). The assumption that parents-to-be are preoccupied with very young children and familiar with the toy symbol could explain the results found. However correct this interpretation might be, it does not explain why height is affected and weight is not.

It is possible that the differences are caused by the nature of the warnings themselves. Respondents know or understand that it can be dangerous if children above a certain weight use certain products (pushchairs, baby-rockers), whilst this relation is less clear with regard to the height of children.

The available data cannot provide an answer to these questions. New tests will have to be conducted to see if the results can be repeated. In these tests, information regarding prior knowledge about the warnings and familiarity with them will have to be collected.

### 13.6  Conclusions

The investigation has resulted in symbols with a sufficiently high level of comprehension for the referents suffocation (92 per cent), unattended (84 per cent) and weight (98 per cent). No symbols with an acceptable level of comprehension were found for age and height, and the reasons for this are neither clear nor straightforward. A general conclusion would be that, *before* a test procedure to select or develop a suitable symbol for a warning is started, the feasibility of a particular warning as a message intended for the general public or a specific user group should be verified. This is because the feasibility of a warning message is determined by the ability of the intended users to specify two aspects of a warning: the possible danger involved and the measures one should take to avoid possible danger. If one and/or the other is insufficiently known, this information should be represented in the proposed symbol.

## Acknowledgements

This research was conducted under contract with the Consumer Safety Institute, Amsterdam (supervision Anita Venema) by Paul Mijksenaar Studio for Visual Information, Amsterdam, Centre for Safety Research, Faculty of Social Sciences, Leiden University (Simone Akerboom and Monica Trommelen), and Psychonomics Department, Utrecht University (Harm Zwaga and Jeroen Visser).

## References

AKERBOOM, S. P., MIJKSENAAR, P., TROMMELEN, M., VISSER, J. and ZWAGA, H. J. G. (1995) *Products for Children: Development and Evaluation of Symbols for Warnings*. Consumer Safety Institute, Amsterdam.

BOERSEMA, Th. and ZWAGA, H. J. G. (1989) Selecting comprehensible warning symbols for swimming pool slides. In: *Proceedings of the Human Factors Society 33rd Annual Meeting*. Santa Monica: Human Factors Society, pp. 994–8.

BRUGGER, Ch. (1990) Advances in the international standardisation of public information symbols, *Information Design Journal*, 6, 79–88.

DINGUS, T. A., HATHAWAY, J. A. and HUNN, B. P. (1991) A most critical warning variable: two demonstrations of the powerful effects of costs on warning compliance. In: *Proceedings of the Human Factors Society 35th Annual Meeting*. Santa Monica: Human Factors Society, pp. 1034–8.

DORRIS, A. L., and PURSWELL, J. L. (1978) Human factors in the design of effective product warnings. *Proceedings of the Human Factors Society 22nd Annual Meeting*. Santa Monica: Human Factors Society, pp. 343–6.

EASTERBY, R. S. and HAKIEL, S. R. (1981) Field testing of consumer safety signs: the comprehension of pictorially presented messages, *Applied Ergonomics*, 12, 143–52.

EDWORTHY, J. and ADAMS, A. (1996) *Warning Design: A Research Prospective*. London: Taylor & Francis.

FRIEDMAN, K. (1988) The effect of adding symbols to written warning labels on user behaviour and recall. *Human Factors*, 30, 507–15.

HAGENAAR, R. and TROMMELEN, M. (1992). *A Study on Hazard and Safety Information Accompanying Children's Products*. Report of the Centre for Safety Research, Leiden.

ISO (1995) *Procedures for the Development and Testing of Public Information Symbols, Revision 1995*, ISO9186. Geneva: *International Organisation for Standardisation*.

JAYNES, L. S. and BOLES, D. B. (1990) The effect of symbols on warning compliance. In: *Proceedings of the Human Factors Society 34th Annual Meeting.* Santa Monica: Human Factors Society, pp. 984–7.

LAUGHERY, K. R., VAUBEL, K., YOUNG, S. L., BRELSFORD, J. W. and ROWE, A. L. (1993) Explicitness of consequence information in warnings, *Safety Science*, 16(5/6), 597–614.

OTSUBO, S. M. (1988) A behavioral study of warning labels for consumer products: perceived danger and the use of pictographs. In: *Proceedings of the Human Factors Society 32nd Annual Meeting.* Santa Monica: Human Factors Society, pp. 536–40.

SANDERS, M. S. and MCCORMICK, E. J. (1993) *Human Factors in Engineering and Design.* New York: McGraw-Hill.

TROMMELEN, M. (1994) *Products for Children: Standardisation of Warnings.* Amsterdam: Consumer Safety Institute.

TROMMELEN, M. (1995) *Products for Children: Testing of Warnings in the Netherlands.* Centre for Safety Research, Leiden University, Leiden.

TROMMELEN, M. and AKERBOOM, S. P. (1997) Explicit warnings for child-care products, *Visual Information for Everyday Use: Design and Research Perspectives.* London: Taylor & Francis (forthcoming).

VAUBEL, K. P. and BRELSFORD, J. W., Jr (1991) Product evaluations and injury assessments as related to preferences for explicitness of warnings. In: *Proceedings of the Human Factors Society 35th Annual Meeting.* Santa Monica: Human Factors Society, pp. 1048–52.

WOGALTER, M. S., GODFREY, S. S., FONTENELLE, G. A., DESAULNIERS, D. R., ROTHSTEIN, P. R. and LAUGHERY, K. R. (1987) Effectiveness of warnings, *Human Factors*, 29(5), 599–612.

WOGALTER, M. S., BRELSFORD, J. W., DESAULNIERS, D. R. and LAUGHERY, K. R. (1991) Consumer products warnings: the role of hazard perception, *Journal of Safety Research*, 22, 71–82.

YOUNG, S. L. and WOGALTER, M. S. (1990) Comprehension and memory of instructions manual warnings: conspicuous print and pictorial icons, *Human Factors*, 32, 637–49.

ZWAGA, H. J. G. and BOERSEMA, Th. (1982) Evaluation of a set of graphic symbols, *Applied Ergonomics*, 8, 87–98.

ZWAGA, H. J. G., HOONHOUT, H. C. M. and VAN GEMERDEN, B. (1991) The systematic development of a set of pictographic symbols for warnings and product information. In: Lovesey, E. J. (ed.), *Contemporary Ergonomics 1991.* London: Taylor & Francis.

# CHAPTER FOURTEEN

# Towards consumer product interface design guidelines

JOHN V. H. BONNER

*Institute of Design, Teesside University*

## 14.1 Are Guidelines Useful?

Product designers are continually confronted with constraints, standards and guidelines. Therefore one response to this chapter would be is there a need for another set of guidelines? Their effectiveness in supporting the design process can certainly be questioned. Research by Klein and Brezovic (1986) asked designers to rate various types of human factors information and found that technical literature was rated least effective, with personal experiences and experimentation supporting design decisions rated the most useful. Designers complained that the literature was often difficult to apply to their particular design problem.

Relevant guidelines are, by their nature, difficult to produce because they have to apply in many situations. Furthermore, guidelines can be axiomatic: statements such as 'present data, messages and prompts in a clear and directly usable form' are self-evident and lack any suggestion or metric by which this statement could be measured. The problem is further compounded by many guidelines assuming that designers only need to be made aware of a rule or principle in order to implement it; or that a few applied psychology principles will provide an adequate design framework.

Designers do not deliberately design bad products. It is more a question of designers not recognising a particular problem existing when provided with design advice. For example, working with the statement 'reduce the amount of memorisation required to complete a command', many designers may critically examine their design with impunity, not recognising that the command sequence may require amendment because they have a different mental model of the users' capabilities. This may be owing to the designer's knowledge domain of product interaction being incomplete at different levels. At a low level, a designer may recognise that providing meaningful and appropriate prompt messages is important but, on the other hand, not recognise that feedback is important when mistakes are made, for example some VCRs do not even beep when a mistake is made (Thimbleby, 1991). At a high level, the designer's understanding of how the product may be used can often be different from the users' or, put another way, the designer's conceptual model of a product may be at variance with the users' as people form mental models of a device through experience, training and formal and informal instruction (Norman, 1988).

When designers are faced with a range of guidelines, they may refer only to topic areas considered applicable and not consider other relevant information. Indeed, it could be argued that the interaction process between a product and a user can be so 'complicated and variable that it is neither possible nor desirable to develop general interface design principles' (Diaper and Schithi, 1995).

However, to argue a complete abandonment of guidelines would be unwise. The implementation of guidelines does require care and consideration, but designers need some form of 'good practice' to assist decision making. Empirical evidence from the field of human factors can be important in guiding and supporting the product development process if designers can readily access this information. Historically, however, consumer products have traditionally been too diverse in application and lacking in complexity to require interface standardisation; but with the increasing use of the microchip, there is a lack of 'interface conventions' in consumer products (Angiolillo, 1995). Consumer products are undertaking a radical change in terms of their functionality, control and management, with increasing 'intelligence' being built into products. Therefore, control and display technologies should be under frequent and critical review (Muckler, 1984). For this reason, product designers need human factors data to design these increasingly more complex products so that they are acceptable and usable by the user.

In contrast, computer-based technologies have developed under common human factors standards (Stewart, 1995), although Pheasant (1987) provides many relevant British standards that designers may find useful. The product designer will find that ergonomics or human factors data and research findings in the area of product interface design are sporadic, and product designers cannot be criticised for not finding relevant information. One of the main objectives of this chapter is, therefore, to provide 'access' to human factors literature that may prove useful in the product development process.

## 14.2  Guideline Parameters

To overcome some of the problems associated with using guidelines, the following guidelines have been structured to address many of the issues raised.

They are written with a checklist format with references to other publications for further reading and information, allowing the designer to source additional material. The guidelines are written to raise awareness of human factors issues and explain where further information can be sought. The number of guidelines is deliberately small in order to address broad issues at a high level of abstraction from the interaction process. Also implicit in all the guidelines is a user centred approach where emphasis is placed on 'performance benchmarking' in user/product interaction.

Terms used in the guidelines are illustrated in Figure 14.1 and have distinct relationships. The product interface refers to all the control, display and feedback mechanisms that exist on a product but are not necessarily physically located together on the product. Interaction is defined as the dialogue between the user and product, and the system is defined as the product/user interaction in the context of a task being performed in certain environmental conditions. The following points should also be noted. The guidelines are:

- based on 'conventional' products and do not include advanced interfaces such as 'intelligent' products;
- general and broad in nature, the regurgitation of specific prescriptive or declarative information from other guideline sources has been avoided;

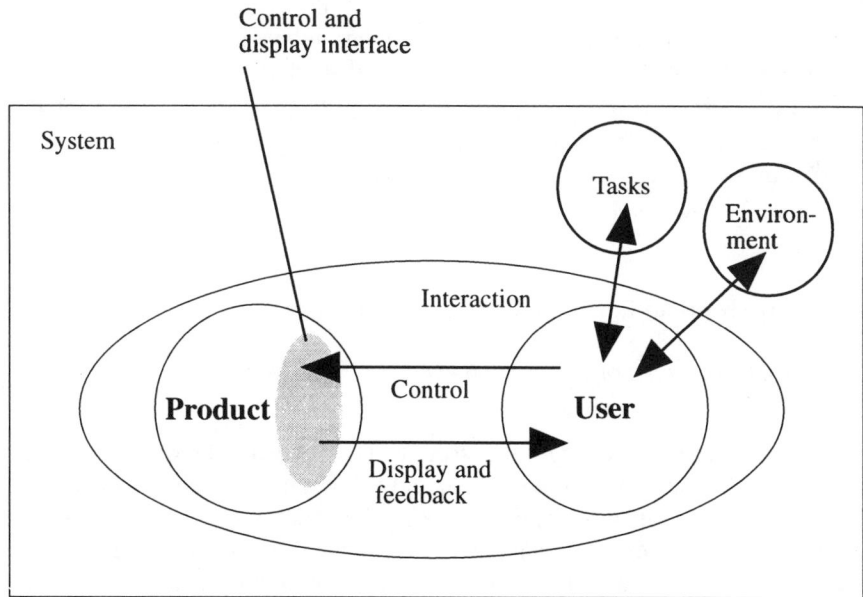

**Figure 14.1**   Interaction design in context.

- from the field of human/computer interaction (HCI) and have obviously been written with the computer user rather than a consumer user, and this should be borne in mind;
- written to be applicable to a wide range of consumer products and are application-independent.

## 14.3  Guidelines

These basic guidelines for good design practice for interface design are based upon a number of sources, predominantly from the fields of human/computer interaction (Asfour *et al.*, 1991; Brown, 1988; Galitz, 1993; Helander, 1990; Nielsen and Molich, 1990; Ravden and Johnson, 1989; Shneiderman, 1992; Smith and Mosier, 1986) and ergonomics texts (Cushman and Rosenberg, 1991; Pheasant, 1987; Salvendy, 1987; Sanders and McCormick, 1993; Woodson *et al.*, 1992). They are grouped into four main sections:

- *Appropriate contextual fit for the interface*   These guidelines discuss the importance of ensuring that the interface is suitable for the relevant user population and can accommodate their initial and dynamically changing requirements, the task or tasks that have to be performed, and the environment in which the interface will be used.
- *Appropriate display devices and feedback*   These guidelines concentrate on designing the interface so that the user has relevant information on their actions, product state changes, guidance and support.
- *Appropriate control of the product state and functions*   These guidelines focus on ensuring that the user is provided with adequate control mechanisms in order to carry out the required task effectively.
- *Adopting a user-centred design approach.*

### 14.3.1  Appropriate Contextual Fit for the Interface

*Ensure that Task Functions Are Matched Correctly between the User and the Product*

As consumer products change, the issues of how, when, who or what activates or controls a function become increasingly difficult to specify. Allocation of function can be defined as explicitly stating the functions between the user and the product. Humans and machines have very different skills and abilities. Decisions about who or what should perform a function should take these skills into account. For example, humans are good at discriminating signals from high levels of visual or auditory noise and making complex decisions using past experience, whereas machines are good at counting or measuring physical quantities and for literal reproduction (Brown, 1988).

However, these distinctions are very broad and do not help more subtle design decisions. As products become more complex in terms of functionality, this distinction becomes more blurred and the designer will have to resort to experiments to determine these factors more precisely. For example, advanced photocopiers take some of the copying decisions away from the user by preventing copying if the original overlaps the designated format size. In this example the system function has been increased by allowing the machine to make copying decisions. This function can be overridden by keeping the 'copy' button depressed. Many users, however, are unaware of this because the communication concerning the transfer of function between the machine and the user has been poorly considered by the designer. A good reference for this area is Kantowitz and Sorkin (1987).

*Ensure the Interface Communicates in a Consistent Manner with the User*

Much of the design literature states that interfaces should be consistent. It is relatively easy to evaluate the consistency of interface elements by listing all the interaction states and noting different descriptive attributes. Ravden and Johnson (1989) present a series of checklists on consistency and other usability issues which may be useful. In summary though, the consistency of the following interface elements should be addressed:

- Ensure all coloured elements have the same meaning throughout the user/ product interaction.
- Text and numerical information should be written in the same style and using the same terms.
- Ensure that all graphical elements are understood by users in different contexts and by different user groups.
- Ensure that instructions do not contradict each other (see Cooper and Page, 1989).
- Control and display layouts should be closely mapped to the task requirements (some form of task analysis will have to be undertaken to achieve this).
- Data entry tasks should use the same protocols, although dangerous or permanent product state changes should require further confirmation or be difficult to process.
- Product interface display and feedback mechanisms should be consistent and appropriate for all stages of the interaction process.

Checking that the interface is consistent with user expectations is more difficult but nevertheless important. It is possible to design a product interface that fulfils all the above criteria which is still inconsistent.

The following considerations are related to other, more broad user expectations that have to be taken into account:

- Ensure compatibility with other products that users may have experience of or would expect to function in a similar way.

- Ensure that the product does not counteract other tasks that may be associated with the product.

- Consistency with the user's mental model of the product, task or relevant experiences.

- Consistency with user stereotypes (many of these are discussed in more detail later).

## Design the Product To Fit Normal Behavioural Patterns

Products are not designed intentionally for the wrong user population. If a product fails to be usable and acceptable, this may be owing in part to an incompatibility between the designer's and user's mental models of the product. It is a common failing to assume that everyone's experiences are the same as one's own. Product designers should therefore develop a comprehensive picture of the different types of user/product interactions that may exist and ensure that the product design allows for this diversity, Loopik *et al.* (1994) provides an example of users' cognitive problems with vacuum cleaners. A useful background to human performance, limitations and differences in systems design can be found in Bailey (1982), whilst Gardner (1987) presents a series of psychological principles which can be applied to interface design.

During product design, the consideration of behavioural activities will inevitably have to be generalised and, to some extent, there are many product/user interactions which can be predicted. Humans are goal-oriented and use artefacts to achieve these goals. This means that users will be task-centred and will find any interface that prevents these objectives frustrating. To ameliorate this, an interface should have certain qualities. Galitz (1993) suggests that they should be:

- adaptive to different users, tasks and environments;

- transparent, allowing the user to be able to concentrate on the task not the product;

- comprehensible, with the user knowing what to do, when to do it and how to do it;

- natural and predictable according to users' expectations;

- responsive, by supplying feedback to user requests;

- self-explanatory, by providing interaction stages that are obvious or supported by appropriate instructions;

- forgiving, by allowing the user to explore and make mistakes;
- efficient, by only providing pertinent information;
- flexible for different user requirements.

To ensure that these qualities are considered, bear in mind the following guidelines in this section.

*Both the Physical Interface and the Interaction Processes Should Adopt or Reflect Familiar Conceptual Models Held by the User*

The designer communicates to the user through the 'system image' (Norman, 1988) which is based on how the designer thinks the system should work. This has also been defined as the 'metacommunication' (Van der Veer *et al.*, 1985) which refers to all activities which communicate the underlying conceptual model of the product. The user's mental model is developed from how the system suggests it works using prior knowledge and experience. For example, students are familiar with an educational model of assignments, lectures and assessments; these metaphors could be used for an educational product aimed at this user population.

*Reduce the Amount of Information the User must Remember in order to use the Interface*

This can be done by providing, for example, default settings, presenting sequences of actions in analogous form, providing clear signposting and navigation through menu structures and keeping categories or classification structures in single figures. Research by Sorce *et al.* (1990) on an information retrieval system for a video library found that subjects preferred to classify video tapes into nine categories as opposed to the 29 found in video shops. Cognitive loading can be reduced by providing the right type and level of functionality to suit the task in hand. Many products offer a wide range of functions which are perceived as being useful by buyers (Lee *et al.*, 1991). However, research suggests that very few of these functions are used because they are difficult to use or remember (Thimbleby, 1991).

*Allow the User To Make Mistakes*

A key rule in human/computer interaction is always to allow the user to undo a command. This cannot always be achieved in consumer products as they tend to have more actions that have a permanent and irretrievable effect, for example selecting record on a tape recorder. Therefore, associated cues or warnings should be provided to prevent accidental error but which allow the user to learn by experimentation and provide clearly marked exits.

## Conform to Stereotypical Behaviour

A stereotype belonging to a particular population can be defined as a commonly expected relationship between a control action or display and the resultant product action. For example, most people assume that green means start and expect that when a control knob is turned clockwise, it will move an object from left to right. However, although these stereotypical actions are widely adopted, some caution must be exercised. Stereotypical behaviour is dependent on experience, which is often culturally related. For example, Courtney (1992) found that Chinese subjects did not have the same colour associations as the western population: red for stop and green for go were not so strongly associated.

Generally, when control-display arrangements conform to population stereotypes there are distinct advantages. Asfour *et al.* (1991) found that reaction or decisions times are shorter, the first control movement made by the user is more likely to be correct and learning times are reduced.

## Accommodate Different Types of Users

User requirements will vary substantially depending on the user populations that they belong to, including:

- children;
- the elderly;
- users with special needs.

Some products such as washing machines require anti-tamper devices to prevent accidental or mischievous use by children. Whilst some products are designed specifically for their use, there is not a significant body of knowledge about characteristics and abilities of children which can be translated into design criteria (Wilson and Norris, 1993). Some development work on a communication tool, however, called PenPal (Piernot *et al.*, 1995) suggests that children prefer products with physically manipulable interfaces and a 'fuzzy boundary' between the physical and software elements of the product. Products should be fun to use and it should be possible to customise the interface.

Products have traditionally been designed for use by people during active adulthood, although there is increasing recognition of the elderly user (Benktzon, 1993; Sandhu, 1993; Vanderheiden 1991). Elderly users require a different type of interaction support. Research by Loring (1993) investigated an elderly user group and their opinions on consumer products and found that they liked to have instructions to learn how to use the product although they complained about the clarity of the instruction material. She also found that the size, number and spacing of the buttons greatly affected ease of use.

Ward (1989) found that for most disabled users, it is the variety of controls found on the product that cause concern. Furthermore, if one key control is inoperable by a user, the product may be unusable for that person. A preferred method for designing for the disabled is to use less variety in the selection of controls on each product. A considerable body of research has been conducted on telecommunication products (von Tetzchner, 1991) and some of these guidelines can be generalised to a wider range of products; Kanis (1993) provides some design recommendations on the controls of consumer products for physically impaired users, and Shein *et al.* (1995) offer advice for the design of computer interfaces for people with physical disabilities.

Other factors that determine differences in aptitude are: experience, both personal and culturally; 'technical aptitude', defined as spatial and reasoning aptitudes and task-specific knowledge. Egan (1988) has, however, argued that other personality-based characteristics are weak and inconsistent predictors of performance.

### The Product Interface Should Accommodate for the Changing Needs and Skill Development of the User

Designing a product interface that is easy to learn can be counterproductive when users become more proficient with the product or related tasks. A step-by-step introductory procedure may become annoying and frustrating to a skilled user. The fundamental difference between a novice and expert user is in their level of knowledge and experience of the system. Therefore ensure that the interface permits different routes and paths to the same goal, including short-cuts. Skilled users prefer a higher level of control of the system.

Gaining knowledge about a system is an important factor in how the user will want to use a product. Shneiderman (1992) divides knowledge into two levels: syntactic and semantic. Syntactic knowledge includes device-dependent information such as which button will cancel a particular function. For users of consumer products, the retention of this type of knowledge is becoming increasingly more difficult as products provide more 'abstract' and similar or hidden functionality. Users may experience confusion when, for example, the remote control of a VCR is different in terms of key layout and symbols from a functionally similar product such as music centre remote control. Many products lack explicitness in conveying the purpose and consequence of control and display devices.

Semantic knowledge refers to concepts that allow the user to represent mentally what is going on inside the product. This is also referred to as the mental model that the user has of the system. Shneiderman (1992) differentiates between users who have a high knowledge of the task but not of the product and vice versa. It is important that the user develops a correct mental model of the product's operations to assist in the learning and retention of its functions.

This picture of changing or varying user needs is further complicated when intermittent or discretionary users, who tend to exhibit characteristics of both novice and expert users, are considered (Santhanam and Wiedenbeck, 1993).

## Ensure that the Product Interface is Understood and Accepted by Different Cultural Groups

As products reach wider global markets this aspect of product interface design becomes more important. Nielsen (1990) provides a useful information resource; although these issues are specifically related to computer-based interface design, they are relevant to product design. Russo and Boor (1993) provide a cross-cultural checklist to ensure that a product bridges cultural boundaries; it includes:

- Translators of text should be aware of the subtleties of product text and should have a full understanding of the product and its use to ensure that some meanings are not misinterpreted. Jargon should not be used. Designers should be aware of other character sets, e.g. Â, Ê, Å, å, ß and ensure these are feasible on electronic displays.

- Beware of the use of commas and full stops to separate large numbers, these vary particularly between Europe and the USA.

- Images and symbols can be misinterpreted across different international markets.

- The interpretation of colour varies widely across countries.

- The placement and arrangement of text and graphical images on the interface should consider other cultures, particularly Arabic and Chinese populations as they read and present information differently.

## Design the Interface To Fit and Support the Intended Tasks

To do this requires a full understanding of the task or tasks. This means that some form of task analysis should be undertaken. Task analysis, in this context, involves the study of user activity to achieve a particular goal. There are many task analyses and many of these are comprehensively described by Kirwan and Ainsworth (1992) and Stammers et al. (1990).

It must be acknowledged, however, that using or selecting a particular task analysis method can be difficult. Benyon (1992) cites various problems. For example, there is a complex array of terms that vary from one task analysis method to another, such as plans, goals, methods, operations, actions. Even the definition of task is questionable. Because there are many approaches to task analysis, a product designer may find it difficult to identify one that will map onto the product design process and provide pertinent results; although Baber and Stanton (1994) offer a synthesised approach using two techniques known as hierarchical task analysis and state space diagrams which highlights where errors and confusion may occur in using an existing or proposed product.

The importance of a designer having a full understanding of the tasks involved in using a product cannot be overstressed, as such an understanding will inevitably facilitate the reduction of usability problems. This requires some rigorous and methodical analysis of the tasks to be undertaken between the user and the product which cannot be explained in any depth here. Instead, some simple advice is presented which can be used in most design scenarios:

- Try to group functionally related devices together; this is typically done by placing controls or display information in close proximity. Other coding methods which have value include colour or shape coding.

- Group frequently used control devices together but also ensure that they are easy to execute and do not interfere with each other.

- Make important, initiation and shut down functions easy to learn.

- Anticipate human errors and design the interface to allow them to happen. For example, provide undo features or security devices to prevent errors. Predicting human error is very unreliable and it is wiser to make observations of typical user/product interactions. In doing this, important lessons can be learnt about the design of the interface. For example, Verhoef (1988) observed users of ticket vending machines and noted the following types of error: information presented by the system may not be noticed by the user; users may not have all the information to hand to complete a transaction; the system may not clearly explain the tasks; or the user may not perform the tasks in the correct order.

*Accommodate Different Environments in which the Product Interface May Be Used*

Sanders and McCormick (1993) and Galer (1987) provide information on environmental issues as well as being good general texts on human factors. Environmental issues that may be relevant and affect the use, and therefore the design, of an interface can be divided into two broad areas: physiological and psychological.

The physiological factors that may affect the usability of an interface, and which, therefore, should be considered, are noise, illumination, climate and motion. Psychological factors include stress, social interaction and mental workload. If these factors are ignored and are not included in the interface design specification, there may be situations where the product may be dangerous or difficult to use.

Noise can be defined as unwanted sound, and it is important that any auditory displays are designed to be discernible not only from the environmental noise but also from each other. Ensure that display and control panels do not potentially present glare that may inhibit the use of the panel, and consider how product interfaces can used in extremely bright or light environments. Some products may be used in extreme climatic conditions. If this is so, ensure that controls can be operated whilst wearing gloves in cold conditions and that the panel does not become too hot to touch in hot conditions. Some products are used in unstable conditions, for example a car radio, where fine adjustments can be difficult – the environmental factor of instability would therefore need to be considered.

Psychological factors that may need to be considered in the design of the product interface include:

- Stress – products such as burglar alarm control panels are sometimes used in stressful conditions. People under stress are accident-prone and tend to revert to rehearsed and stereotypical behaviour.

- Consider that products may used while the user is in a 'transitory state' such as tired, fatigued, under the influence of alcohol or drugs.

- Social and cultural factors may determine the usage of a product. Crozier (1994) discusses psychological responses to design by addressing issues such as the meaning and emotional response to products.

### 14.3.2  Product Presentation

*The Product Interface Should Inform How and When It Should Be Used through Various Display Mechanisms*

A product display can be online, either dynamic (various forms of instruments that vary with time) or a static display information which does not change with time. Offline material includes manuals and operating instructions. The following types of display should be considered:

- *Alphanumeric displays* can be used for a variety of purposes such as instructions, prompts or labelling. There is plenty of literature on this subject, particularly on character recognition, height-to-width ratio, stroke width, letterspacing between and within words, spacing between lines and viewing distances. See, for example, Cushman and Rosenberg (1991) and Woodson *et al.* (1992) for detailed information. A good publication on writing instructions for consumer products is Cooper and Page (1989). There is very little literature on the effectiveness of instructions for specific product types.

- *Colour* can be used to highlight, group and code different interface objects (Van Laar and Flavell, 1990). These should, as far as possible, comply with standards or *de facto* standards. (See Diaper and Schithi, 1995, for some controversial views on this subject which relate to more general design issues on guidelines.)

- *Graphics* is defined here in its broadest terms, and can be used to display complex information, trends, predictive versus actual values and dynamic information. Graphic displays include symbols (where Easterby (1970) states that symbols should fulfil two major criteria of discriminability and meaning) and icons, which are small graphic images that represent familiar objects or abstract forms in order to convey their function. See Galitz (1993), Gittens (1986) and Lin *et al.* (1992) for specific guidelines and design issues.

- *Auditory displays* should be used when visual displays are not appropriate (see Blattner *et al.*, 1989; Gaver, 1989), for example when driving a car and trying to tune in the radio or when there is a need to reduce the cognitive load on the visual channel. Sound has the advantage that it can be heard from all directions and the user does not need to be focusing on the output device. Auditory displays should be used to enhance visual displays and make their operation more memorable and when visually impaired users need to use a visual based interface. Reaction to auditory stimuli is faster than reaction to visual stimuli (Bly, 1982).

- Form or shape of controls can display how a control should be used, Norman (1988) refers to this as 'affordance'. For example, washing machine control knobs have been shaped in the form of an 'S' to suggest the direction the control should be turned.

- Dials and pointers are rarely used on products and therefore not discussed in detail here, though many general ergonomics books have detailed information if it is required.

- Grouping and clustering of displays can provide important cues on the functional process of the product. Spatial considerations should be closely linked with task requirements.

- Electronic displays are increasingly being used on consumer products. The field of human/computer interaction has much to offer in terms of guidelines that could be used in display design. Cushman and Rosenberg (1991) offer useful advice on selecting an appropriate display technology. Galitz (1993) is an extremely comprehensive book on screen design discussing user issues, the use of different dialogue styles, graphical screens and colour.

- Indicator lights are a useful display medium. Indicator lights can be used to indicate whether power is on or off, the status of a function, mode or hazard conditions or that a malfunction has taken place. See Woodson *et al.* (1992) and Pheasant (1987) for uses and effectiveness of indicator light displays. In contrast to industrial products, where coloured indicator lights have recognised meanings, consumer products do not comply so rigorously to standards.

- Conspicuity of target objects. Cole and Hughes (1984) define two types of conspicuity: attention conspicuity measures the propensity of a display object to attract attention when the observer's attention is elsewhere, whilst search conspicuity measures the likelihood of a display object being located during a search for that object.

- Coding should be used with care, avoiding a myriad of display coding methods. Coding types include colour, text and letters, geometric shapes, size, brightness and flash rates.

- Semantic cues – the conscious use of product form and visual cues or metaphor to optimise the interaction between the product and user – can be provided. There are three important elements that should be considered when presenting a semantic cue: what the semantic cue(s) should be; how the representation should be conveyed (for example, visual or auditory); and how the semantic cues are interpreted by the user.

*Ensure that the Product Provides Effective Control Mechanisms which Are Understood*

Guidelines and information on control devices are numerous, although many of the control devices cited are not relevant to modern consumer products. Hard control devices (as opposed to soft control devices such as touchscreens) are liked by consumers and there is ample scope to develop these types of device further (Black and Buur, 1996). There have been few studies that determine consumer preferences for controlling consumer products. However, a study by Sorce *et al.* (1990) examined user preferences for a video service and found that subjects preferred a hand-held controller to a keyboard or joystick, a TV screen to a laptop computer screen for selecting a show, and a visual rather than an audible dialogue with the system. In a study of car radios (Johnson, 1991), subjects preferred rotary control knobs over other types (which were not specified in the paper) and preferred rotary knobs with resistance within the control.

The variation of control devices for consumer products is surprisingly narrow. A recent study by Bonner (1995), noted that 89 per cent of controls were pushbutton and that, generally, users found them too small. Similarly, Sandhu (1993) found that, for elderly users, control knobs rarely conform to ergonomics criteria. They are often too small, difficult to turn, provide little feedback and are too close to each other. In designing pushbuttons, the following criteria should be taken into account: dimensions, shape, grouping, guarding, forces and profile. For further information on control design, see Pheasant (1987).

### 14.3.3   Providing Appropriate Feedback

*Make Sure the User Knows What Is Going On*

Appropriate feedback in the context of consumer product design can be defined as conveying to the user the right level of information before, during and after the execution of an action. Norman (1992) suggests that a person needs at least three different kinds of confirming information: the act itself, intermediate results and a final outcome. Shneiderman (1992), describes this process as closure, where actions have a beginning, a middle and an end. That is to say that, at each stage of the interaction process, it should be apparent to the user where to begin, it should be apparent that the product has undertaken an instruction and is carrying out the task, and it should be apparent when this is complete.

The feedback process is complex in that it is dynamic and multimodal, making it difficult to measure the effectiveness by which users understand different feedback mechanisms. As product/user interactions become more abstract and removed from physical actions, feedback mechanisms need to be deliberately designed into the product. The only way to ensure that adequate feedback has been satisfactorily designed into the interaction process is to conduct user evaluations. The following types of issues should be addressed and considered in an evaluation programme:

- Assess whether the feedback is offering informative feedback: the product should only tell the user what he/she needs to know in order to continue work productively. It is important to find out if users obtain the right type of feedback.
- Check that the user has correctly understood the feedback provided.
- Make sure that the user is aware of state and mode changes and adequate support is provided when things go wrong. Conn (1995) provides a detailed study of response time particularly related to interactive systems and provides some useful principles.
- Navigation signposts should be given; for example, attempts should be made to present all relevant information to enable an action to be completed. Users should not have to remember data from one display page to another (Engel and Granda, 1975).
- Evaluate the help systems and ensure that they provide relevant and timely support.
- Warning displays can be either active, by presenting warning information when it is relevant, or passive such as labels and signs. See Silver and Braun (1993) for readability issues of warning labels, and Frantz (1993) for information on the location and presentation of warning instructions. Lehto (1991) states that it is important to match the warning design to the operator's level of performance. Some form of task analysis should take place to identify where errors may occur and, subsequently, a product evaluation conducted to establish whether the warnings are discernible, correctly perceived and interpreted.

### 14.3.4  Work Directly with Users

Including end-users in the design process is not a trivial consideration. From an economic, organisational or practical point of view, it can be difficult to implement. However, if this approach is considered then users can be involved in several ways:

- As part of the design team (participative design), for example in focus groups to develop solutions to product problems (Caplan, 1990).

- In user trials for existing or prototype products where testing user behaviour is measured rather than just opinions. McClelland (1990) gives a very good description of user trials. However, as products become more interactive it is the 'interactivity' that needs to be designed rather than the physical product (Webb, 1996).

- In formulating user requirements in terms of product needs.

- In developing a user profile which describes the cultural, physiological and psychological aspects of the user.

## 14.4 Conclusions

The literature review that was carried out for this chapter has revealed many areas that need further investigation. Firstly, as consumer products begin to converge bringing together computer, telecoms and electronic consumer products, interaction issues will play a larger role in the design process to provide 'usable' products. The need for a greater understanding about user needs becomes more acute with an increased emphasis on participatory design required to accomplish product design objectives (Sanders, 1993). A greater understanding is required of consumers as a user population, particularly in fields where products begin to have more sophisticated interaction styles. Unlike the majority of computer users in an industrial context, consumers make their own purchasing decisions, at which point usability may not be a high priority, and can then subsequently be highly discretionary in using them. This may ultimately result in total abandonment of large parts of product functionality if the product lacks the transparency to allow the user to achieve a required task.

Secondly, in contrast to the commercial context where there is usually formal or informal product knowledge, the training and learning in the consumer market is usually totally dependent upon printed instructional advice, usually without any personal advice or experience to provide additional support. Thirdly, few human factors standards exist on how to design consumer products which may provide consistency in interaction dialogues and therefore reduce learning times and confusion. Finally, further investigation is required on emerging and conceptual interaction styles to assess the usability implications. It is important that products have an opportunity to be led by user needs and characteristics rather than be completely technology-led.

To summarise, the intention of this chapter has been to provide the product designer with some broad guidelines which can also act as a checklist to ensure that ergonomic issues of consumer interfaces, particularly those containing electronic displays, are addressed. References to more detailed recommendations and guidelines have been provided and should be regarded as pointers to more specific and detailed research.

## References

ANGIOLILLO, J. S. (1995) Minimal remote: a standard input device for consumer interactive TV. In *Proceedings of the 39th Annual Meeting of the Human Factors and Ergonomics Society*, pp. 194–7.

ASFOUR, S. S., OMACHONU, V. K., DIAZ, E. L. and ABDEL-MOTY, E. (1991) Displays and controls. In: Mital, A. and Karowski, W. (eds), *Workspace, Equipment and Tool Design*. Amsterdam: Elsevier, pp. 257–76.

BABER, C. and STANTON, N. A. (1994) Task analysis for error identification: a methodology for designing error-tolerant consumer products, *Ergonomics*, **37**, 1923–41.

BAILEY, R. W. (1982) *Human Performance Engineering: A Guide for System Designers*. Englewood Cliffs, NJ: Prentice Hall.

BENKTZON, M. (1993) Design for our future selves: the Swedish experience, *Applied Ergonomics*, **24**, 19–27.

BENYON, D. (1992) The role of task analysis in system design, *Interacting with Computers*, **4**, 102–23.

BLACK, A. and BUUR, J. (1996) Making solid user interfaces work, *Information Design Journal*, **8**(2), 99–108.

BLATTNER, M. M., SUMIKAWA, D. A. and GREENBERG, R. M. (1989) Earcons and icons: their structure and common design principles, *Computer Interaction*, **4**, 11–44.

BLY, S. A. (1982) Presenting information in sound, *CHI: Proceedings of the ACM Conference on Computer–Human Interaction*. New York: ACM, pp. 371–5.

BONNER, J. V. H. (1995) A review of control and display characteristics of contemporary products. Paper presented at Ergonomics in Consumer Product Design and Evaluation Conference, 10 November, University of Southampton.

BROWN, C. M. (1988) *Human–Computer Interface Design Guidelines*. USA: Ablex Publishing.

CAPLAN, S. (1990) Using focus group methodology for ergonomic design, *Ergonomics*, **33**(5), 527–33.

COLE, B. L. and HUGHES, P. K. (1984) A field trial of attention and search conspicuity, *Human Factors*, **26**(3) 299–313.

CONN, A. P. (1995) Time affordances: the time factor in diagnostic usability heuristics. In: Katz, I. R. *et al.* (eds), *CHI'95: Mosaic of Creativity, Proceedings of Conference on Human Factors in Computer Systems*, pp. 186–93.

COOPER, S. and PAGE, M. (1989) *Instructions for Consumer Products*. London: Department for Trade and Industry.

COURTNEY, A. J. (1992) Control: display stereotypes for multicultural user systems, *IEEE Transactions on Systems, Man, and Cybernetics*, **22**(4), 681–7.

CROZIER, R. (1994) *Manufactured Pleasures: Psychological Responses to Design*. Manchester: Manchester University Press.

CUSHMAN, W. H. and ROSENBERG, D. J. (1991) *Human Factors in Product Design*. Amsterdam: Elsevier.

DIAPER, D. and SCHITHI, P. S. (1995) Red faces over user interfaces: what should colours be used for? In: Kirby, M.A.R. *et al.* (eds), *People and Computers X*. Cambridge: Cambridge University Press, pp. 425–35.

EASTERBY, R. S. (1970) The perception of symbols for machine displays, *Ergonomics*, **13**(1), 149–58.

EGAN, D. E. (1988) Individual differences in human–computer interaction. In: Helander, M. (ed.), *Handbook of Human–Computer Interaction*. Amsterdam: North Holland.

ENGEL, S. E. and GRANDA, R. E. (1975) *Guidelines for Man/Display Interfaces*, Technical Report TR 00.2720, IBM, Poughkeepsie, New York.

FRANTZ, J. P. (1993) Effect of location and presentation format on attention to and compliance with product warnings and instructions, *Journal of Safety Research*, **24** (3), 131–54.

GALITZ, W. O. (1993) *User Interface Screen Design*. Boston: QED Information Sciences.

GARDNER, C. (1987) Tumbling back from the brink, *Design*, **461**, 28–9.

GALER, I. A. R. (ed.) (1987) *Applied Ergonomics Handbook* (2nd edition). London: Butterworth.

GAVER, W. W. (1989) The Sonic Finder: an interface that uses auditory icons, *Human/Computer Interaction*, **4**, 67–94.

GITTENS, D. (1986) Icon-based human–computer interaction, *International Journal of Man/Machine Studies*, **24**, 519–43.

GOLDSMITH, S. (1984) *Designing for the Disabled* (3rd edition). London: RIBA.

HELANDER, M. (ed.) (1990) *Handbook of Human–Computer Interaction*. Amsterdam: North Holland.

JOHNSON, J. (1991) Modes in non-computer devices, *International Journal of Man/Machine Studies*, **32**, 423–38.

KANIS, H. (1993) Operation of controls on consumer products by physically impaired users, *Human Factors*, **35** (2), 305–28.

KANTOWITZ, H. H. and SORKIN, R. D. (1987) Allocation of function. In: Salvendy, G. (ed.), *Handbook of Human Factors*. New York: Wiley, pp. 355–69.

KIRWIN, B. and AINSWORTH, L. K. (1992) *A Guide to Task Analysis*. London: Taylor & Francis.

KLEIN, G. A. and BREZOVIC, C. P. (1986) Design engineers and the design process: decision strategies and human factors literature. In: *Proceedings of the Human Factors Society 30th Annual Meeting*, pp. 771–5.

LEE, M. W., YUN, M. H., PARK, D., CHUN, Y. H., JUNG, E. S. and FREIVALDS, A. (1991) E. Y. E. S. – Ergonomics in a conceptual design process for consumer electronic products, *Human Factors Society 35th Annual Meeting, San Francisco, California*, vol. 1, pp. 466–70.

LEHTO, M. R. (1991) A proposed conceptual model of human behaviour and its implications for design of warnings, *Perceptual and Motor Skills*, **73** (2), 595–611.

LIN, R., KREIFELDT, J. G. and CHI, C-F. (1992) A study of evaluation design sufficiency for iconic interface design perspective. In: Lovesey, E.J. (ed.), *Contemporary Ergonomics*. London: Taylor & Francis, pp. 376–84.

LOOPIK, W. E. C., KANIS, H. and MARINISSEN, A. H. (1994) The operation of new vacuum cleaners: a users' trial. In: Robertson, S. A. (ed.), *Contemporary Ergonomics*. London: Taylor & Francis, pp. 34–9.

LORING, B. A. (1993) Survey of older customers and home electronics: trends and usability issues. In: *Proceedings of Interface '93, Raleigh, North Carolina*, pp. 95–100.

MCCLELLAND, I. (1990) Product assessment and user trials. In: Wilson, J. R. and Corlett, E. N. (eds), *Evaluation of Human Work: A Practical Ergonomics Methodology*. London: Taylor & Francis.

MUCKLER, F. A. (1984) Standards for the design of controls: a case history, *Applied Ergonomics*, **15** (3), 175–8.

NIELSEN, J. (ed.) (1990) *Designing User Interfaces for International Use*. New York: Elsevier.

NIELSEN, J. and MOLICH, R. (1990) Heuristic evaluation of user interfaces. In: *Proceedings of CHI'90 Conference, New York*, pp. 249–56.

NORMAN, D. A. (1988) *The Psychology of Everyday Things*. New York: Basic Books.

NORMAN, D. A. (1992) Design principles for cognitive artefacts, *Research in Engineering Design*, **4** (1), 43–50.

PHEASANT, S. (1987) *Ergonomic Standards and Guidelines for Designers*. Milton Keynes: BSI.

PIERNOT, P. P., FELCIANO, R. M., STANCEL, R., MARSH, J. and YVON, M. (1995) Designing the PenPal: blending hardware and software in a user-interface for children. In: *CHI'95 – Mosaic of Creativity, Proceedings of Conference on Human Factors in Computer Systems, Denver, Colorado*, Katz, I.R. *et al.* (eds), pp. 511–18.

RAVDEN, S. and JOHNSON, G. (1989) *Evaluating Usability of Human–computer Interfaces*. Chichester: Ellis Horwood.

ROUSE, W. B. (1991) *Design for Success: A Human Centred Approach to Designing Successful Production and Systems*. Chichester: Wiley.

RUSSO, P. and BOOR, S. (1993) How fluent is your interface? Designing for international users. In: *Proceedings of INTERCHI 93 Conference on Human Factors in Computing*, pp. 342–7.

SALVENDY, G. (1987) *Handbook of Human Factors*. Chichester: Wiley.

SANDERS, E. B-N. (1993) Converging perspectives in product development research, *Interface 93*, pp. 236–41.

SANDERS, M. S. and MCCORMICK, E. J. (1993) *Human Factors in Engineering and Design* (7th edition). Maidenhead: McGraw-Hill.

SANDHU, J. (1993) Design for the elderly: user-based evaluation studies involving elderly users with special needs, *Applied Ergonomics*, **24** (1), 30–4.

SANTHANAM, R. and WIEDENBECK, S. (1993) Neither novice nor expert: the discretionary user of software, *International Journal of Man-Machine Studies*, **38**, 201–29.

SHEIN, F., ENG, M. and ENG, P. (1995) Access considerations of human–computer interfaces for people with physical disabilities. In: Anzai, Ogawa and Mori (eds), *Symbiosis of Human and Artefact: Future Computing and Design from HCI*, Amsterdam: Elsevier, pp. 143–8.

SHNEIDERMAN, B. (1992) *Designing the User Interface: Strategies for Effective Human/Computer Interaction* (2nd edition). Reading, Mass.: Addison-Wesley.

SILVER, N. C. and BRAUN, C. C. (1993) Perceived readability of warning labels with varied font sizes and styles, *Safety Science*, **16** (5/6), 615–25.

SMITH, S. L. and MOSIER, J. N. (1986) *Guidelines for Designing User Interface Software*, Technical Report ESD-TR-86-27, USAF Electronic Systems Division, Hanscom Air Force Base, Massachusetts.

SORCE, J., FAY, D., RAILA, B. and VIRZI, R. (1990) Designing a broadband residential entertainment service: a case study. In: *Proceedings of the 13th International Symposium, Human Factors in Telecommunications, Turin, Italy*, pp. 141–8.

STAMMERS, R. B., CAREY, M. S. and ASTLEY, J. A. (1990) In: Wilson, J.R. and Corlett, E.N. (eds), *Task Analysis in Evaluation of Human Work: A Practical Guide*. London: Taylor & Francis.

STEWART, T. (1995) Ergonomics standards concerning human–system interaction: visual displays, controls and environmental requirements, *Applied Ergonomics*, **26** (4), 271–4.

THIMBLEBY, H. (1991) Can anyone work the video?, *New Scientist*, 23 February 1991.

VAN DER VEER, G. C., TAUBER, M., WAERN, Y. and VAN MUYLWIJK, B. (1985) On the interaction between system and user characteristics, *Behaviour and Information Technology*, **4** (4), 289–308.

VAN LAAR, D. and FLAVELL, R. (1990) How to use colour in displays: physiology, physics and perception, *ICC Technical Journal*, **7** (1), 154–79.

VANDERHEIDEN, G. C. (1991) Design principles to increase the accessibility of mass market consumer products. In: *Interface '91 Proceedings of the 7th Symposium on Human Factors and Industrial Design in Consumer Products, Dayton, Ohio*, pp. 374–8.

VERHOEF, C. W. M. (1988) Decision making of vending machine users, *Applied Ergonomics*, **19** (2), 103–9.

VON TETZCHNER, S. (ed.) (1991) *Issues in Telecommunication and Disability*, Commission of the European Communities, ECSC-EEC-EAEC, Brussels.

WARD, J. T. (1989) Human factors design guidelines for the disabled. In: *Proceedings of the Human Factors Society 33rd Annual Meeting, Denver, Colorado*, vol. 1, pp. 490–2.

WEBB, B. R. (1996) The role of users in interactive system design: when computers are theatre, do we want the audience to read the script?, *Behaviour and Information Technology*, **15** (2), 76–83.

WILSON, J. R. and NORRIS, B. J. (1993) Knowledge transfer: scattered sources to sceptical clients, *Ergonomics*, **36** (6), 677–86.

WOODSON, W. E., TILLMAN, B. and TILLMAN, P. (1992) *Human Factors in Design Handbook: Information and Guidelines for the Design of Systems, Facilities, Equipment and Products for Human Use* (2nd edition). New York: McGraw-Hill.

# PART FOUR

# Key Topics

# CHAPTER FIFTEEN

# Key topics in consumer products

NEVILLE STANTON

*Department of Psychology, University of Southampton*

## 15.1 Key Topics

This book has covered the diverse topic areas of consumer products from a human factors viewpoint. The contributors come from academia and industry and have a mixture of backgrounds, including psychology, ergonomics and design. The hope is that this text will stimulate further interest, reading, research and development. A content analysis of the contributions has revealed four key topics: ergonomics methods, standards and legislation, user-centred design process, and accidents and errors. This concluding chapter will draw the discussions of the contributors together under these topics.

The range of consumer products addressed in varying degrees throughout this book, as shown in Table 15.1. The examples illustrate tangible contributions that human factors can make to the design of consumer products.

## 15.2 Ergonomics Methods

In their survey, Baber and Mirza (Chapter 5) suggested that only a limited range of methods tend to be used in the evaluation of products, typically focusing on interviews, observation and questionnaires. Stanton and Young (Chapter 2) reinforced this observation and introduce a range of methods that could be applied to consumer product design and evaluation, such as:

- heuristics;
- checklists/guidelines;
- observation;

**Table 15.1** Examples of consumer products within the book

| Consumer product | Chapter |
|---|---|
| Teapot | 1 |
| Desk lamp | 1 |
| Radio cassette | 2 |
| Lawnmower | 3, 7 |
| Hairdryer | 3 |
| Playground equipment | 3 |
| Kettle | 4, 5, 9 |
| Video cassette recorder | 4 |
| Cooker | 5, 6, 9 |
| Washing machine | 5 |
| Freezer | 5 |
| Power tools for DIY | 7, 9 |
| Gardening equipment | 7, 9 |
| Children's equipment | 7, 13 |
| Vacuum cleaner | 9 |
| Telecommunications equipment | 9 |
| Rucksack | 10 |
| Kitchen utensils | 10 |
| Knives | 10, 12 |
| Electric razor | 10 |
| Strimmer | 11 |
| Pen | 12 |
| Scissors | 12 |

- interviews;
- questionnaires;
- link analysis;
- layout analysis;
- hierarchical task analysis;
- SHERPA;
- task analysis for error identification;
- repertory grids;
- keystroke level model.

Each of these methods is briefly discussed and an example of the output provided, with the aim of encouraging people to explore the different methods for themselves. Baber and Mirza (Chapter 5) cite the lack of knowledge and training in these methods as one reason for the people's reluctance to use them. Evans (Chapter 11) reinforced this point.

Hierarchical task analysis (HTA) seems to be a core tool in the ergonomists' repertoire, even if it is a modified version of the technique adapted for the specific purposes of a design team. Jordan *et al.* (Chapter 8) discuss how they have successfully applied a goal-based analysis method (called TAD) in interaction design to assist them in preparing interface specifications. They have also been developing their own methods, such as the RAI, to help them choose between alternative product designs to take accounts of both positive and negative ratings from a sample user group.

Two 'new' methods for consumer product design were introduced. Benedyk and Minister (Chapter 3) introduce the BeSafe method and Stanton and Baber (Chapter 4) introduce task analysis for error identification (TAFEI). It is interesting that both these methods aim to identify possible aberrant behaviour with consumer products. The BeSafe method was developed in the coalmining industry and is based upon the general framework of Reason's (1990) accident causation model. BeSafe takes an auditing approach using a range of ergonomics techniques, but is intended for use by non-ergonomists. The four main phases associated with BeSafe are an ergonomic audit (to identify predisposition to errors), instructional and behavioural audits (to identify predisposition to violations), and a grouping of critical factors (to identify predisposition to latent failures). The outcome of the audit is to identify interventions that will improve product safety. By way of contrast, TAFEI was developed primarily for assessment of consumer products. TAFEI takes a systems approach to scrutinise human/product activity in detail. The method focuses on the transitions between product states brought about by user activity to identify three basic types: legal (error-free transition), illegal (erroneous transitions) and impossible transitions. It is argued that the designer should attempt to reduce the number of illegal transitions by turning them into impossible transitions.

Butters (Chapter 9) discusses a range of methods used in the evaluation of consumer products by the Consumers' Association (CA). It is argued that the choice of method will largely be product-based: home user trials for small products such as kettles and vacuum cleaners; user trials in the CA laboratory for large products such as washing machines and products that require controlled environments such as gardening and DIY; in-house panels for cars, audiovisual and telecommunications products; specialist user groups for software. As a supplement to the user trials, other methods such as checklists, expert appraisals and diary studies are utilized. Similarly, the Institute for Consumer Ergonomics (ICE) uses a combination of observation in the field, focus groups, expert appraisals and user trials (Chapter 7). Evans (Chapter 11) indicates how a range of methods may be combined into the product design cycle. Methods discussed include questionnaires, surveys, user trials, focus groups and simulations. Finally, Bonner (Chapter 14) and Hedge (Chapter 12) offer sets of guidelines which could form part of a design methodology. Bonner's guidelines are divided into four general areas: contextual fit, display and feedback, control of product and functions, and user-centred design. Hedge's guidelines are rather more specific as they relate primarily to hand-operated products.

The range of methods covered in the book is indicated in Table 15.2.

The choice of method will be dependent upon a number of factors. Four that may influence the decision are: the stage of the design process (i.e. early, middle or late) the form that the product takes (i.e. concept, prototype or finished), access to end-users and the degree of time pressure (Chapter 2). Another factor of cost constraints was added by Jordan *et al.* (Chapter 8). Butters (Chapter 9) identifies two further factors: the characteristics of the product under examination and the frequency with which the product is tested. These seven factors should help people to identify the relevant combination of methods for user-based product assessment. Ideally, the final selection of methods should capture a range

**Table 15.2** Range of methods covered in the book

| Method | Chapter |
|---|---|
| Heuristics | 2 |
| Layout analysis | 2 |
| Checklists | 2, 5, 9 |
| Hierarchical task analysis (HTA) | 2, 8 |
| Direct observation | 2, 5, 6, 7, 9 |
| SHERPA | 2 |
| Interviews | 2, 5, 11 |
| Task analysis for error identification | 2, 4 |
| Questionnaires | 2, 5, 11 |
| Repertory grids | 2 |
| Link analysis | 2 |
| Keystroke level model (KLM) | 2 |
| BeSafe | 3 |
| Standards | 1, 5, 6, 13 |
| Expert appraisal | 5, 7, 9 |
| Rating scales | 5, 6, 8, 13 |
| Videotaping | 5 |
| Focus groups | 6, 7, 8, 11 |
| Guidelines | 6 |
| Storyboarding | 6 |
| Simulation | 6, 11 |
| Accident investigations | 7, 11 |
| User trials | 7, 8, 9, 11 |
| Diary studies | 9 |
| Survey | 11 |
| Prototyping | 11 |
| Guidelines | 12, 14 |

of representative user behaviour and should be chosen for divergence (i.e. they measure different aspects of user behaviour) as well as convergence (i.e. they offer some measure of cross-validation). It is argued that both subjective and objective data should be sought from user trials (Chapter 9).

## 15.3 Standards and Legislation

Various standards and legislation were referred to throughout the text, indicating how important these are to product design. Legislation in the form of the Sales of Goods Act was briefly introduced (Chapter 1) and the more recent Sale and Supply of Goods Act (1994) was also discussed (Chapter 7). The General Product Safety Regulations (1994) put a responsibility on the producers and distributors of products to ensure that the goods are safe (Chapters 3 and 7). The Product Liability and the Consumer Protection Act (1987) covers the provision of safety with respect to the marketing of products, instructions, warnings and use of products. Baber and Mirza (Chapter 5) point out that most standards that apply to products do not require ergonomics methods to be undertaken to comply.

The roles of British, European and International standards in consumer products were introduced in the first chapter. ISO 9241 was mentioned several times throughout the book (Chapters 1, 5 and 8). Although intended primarily for computer interfaces, the standard appears to embody most aspects of good design practice which could be applied to the design of products. In any case, so many consumer products contain microchips that the definition of what constitutes a computer-based interface is very broad. The workings of ISO 9186 were illustrated in the development of warning symbols for child-care products (Chapter 13). This approach makes the point of involving would-be users in the development and assessment of symbols. Symbols as warnings have the distinct advantages that they draw attention to the warning – and in doing so, inform the user of a potential danger – and there are no language barriers. Despite these advantages, symbols are notoriously difficult to understand. To counteract the problem of possible miscomprehension, the acceptance criteria are set at a very high level in an attempt to exclude the confusing symbols. It may not be possible to determine a symbol for all possible situations. Trommelen and Zwaga (Chapter 13) argue to increase the effort in determining whether it is possible to construct a meaningful symbol before engaging in exhaustive testing. This would reduce the likelihood of a fruitless search.

## 15.4 User-centred Design Process

All of the chapters in this book have stressed the need for a user-centred focus in design. There is a call for a closer working relationship between the designer and the ergonomist (Chapters 1, 6, 8, 10 and 11). Involvement of the user in design seems to be at several different levels. First is involvement of the user in market surveys and assessment of concepts in focus groups (Chapters 7, 8, 9 and 11). Second is the representation of the user characteristics, such as the structure of the hand (Chapter 12). Third is the use of golden rules about users and products in the form of design guidelines (Chapters 12 and 14). Fourth is the expert appraisal of user activity through ergonomics methods (Chapters 2, 3, 4 and 5). Fifth is the involvement of the user in simulated trials with prototypes and real trials with working products (Chapters 7 and 11). Finally, is the involvement of users in product evaluation activities after product release (Chapters 7 and 9). The degree to which user involvement has been part of the design philosophy is likely to be a large factor in the final usability of the product.

The involvement of users in product evaluations is a matter to be treated with some care (Chapter 9). A product evaluation starts by defining the user population and their characteristics (such as age range and any special needs). Users should be carefully briefed before the beginning of the study and debriefed when the trial is over. There are ethical guidelines for the involvement of participants in studies; these are available in the UK from the British Psychological Society (tel. +44 (0) 116–254 9568 or fax +44 (0) 116–247 0787 or email: bps1@le.ac.uk). It is also advised that participants are not overloaded. Butters (Chapter 9) recommends no more than three hours should be spent in a laboratory study and a maximum of five weeks on a home trial.

Although variations on the user-centred design process have been portrayed throughout the book (Chapters 1, 6, 8 and 11), there are many commonalities in the stages. Variations are normally about how many iterations of the stages are included for the design of any one product. This may be more a factor of degree, i.e. whether the new product represents only a minor incremental change on its predecessors or is a completely new product. More user involvement will be required with the latter option. Another factor would be the proportion of budget allocated to user involvement in the development of the product.

The issue of aesthetics is a topic rarely mentioned in human factors, and it has been tackled by a designer in this book. Macdonald (Chapter 10) argues that, in understanding the user, we need to explore beyond the traditional human factors concerns to understand how we perceive pleasure. He calls for human factors to consider the qualitative experience of users, such as personal values, feelings, self-image and sensation. Ergonomists are beginning to awaken to these ideas and are developing measures to capture the users' experience with products. Certainly, many of the measures taken in user trials require participants to offer subjective judgements, such as in interviews, questionnaires and repertory grids. Although Macdonald voices some concern about possible boundaries between human factors and aesthetics, there may be more in common than readily recognised by both communities. Certainly the design principles encapsulated by the Bauhaus movement, such as simplicity of

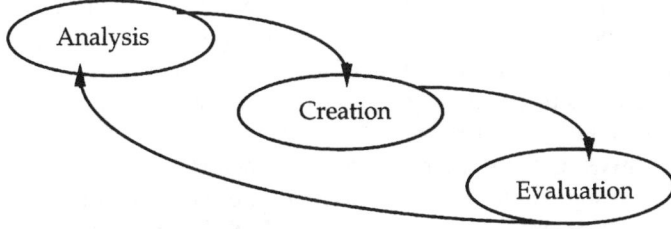

**Figure 15.1**   Generic design phases.

form and form following function, seem to coincide with what would be considered good human factors. Evans (Chapter 11) argues that industrial designers possess the necessary skills to complement the science of ergonomics.

In summary, user-centred design does not simply mean asking users what they think (focus groups, interviews), but designing with user capacities and abilities in mind (guidelines and ergonomics methods) as well as assessing the performance of users with products (user trials). Jordan *et al.* (Chapter 8) identify the three generic phases in product design as analysis, creation and evaluation. These are illustrated in Figure 15.1. As the figure shows, feedback between the phases can help to inform the designer of appropriate design modifications.

Hedge (Chapter 12) points out that design of products needs to capture four basic characteristics of the user/product relationship: characteristics of the product user, characteristics of the environment in which the product is to be used, characteristics of the task to be performed, and characteristics of the product. Truly effective design will require a multidisciplinary approach incorporating designers, engineers, material scientists, software engineers, ergonomists, accountants and so on. Each discipline has a unique contribution to make to the design process.

## 15.5   Accidents and Errors

It is not surprising that accidents and errors feature so largely in a book of this nature. Baber and Mirza (Chapter 5) point out that about 250 000 people a year in the UK require hospital treatment for accidents that occur in the kitchen. It is estimated that 60 000 non-fatal accidents in the home result from the use (or misuse) of consumer products (Chapter 7). Hedge (Chapter 12) points out that the failure to incorporate good ergonomics in the design of products could lead to an increase in the risk of injury. From BS 3456 (Chapter 1) it seems that product designers need not only to assess normal use of their products, but also to look at potential abnormal and careless usage. The human error prediction methodologies might be useful in this endeavour (see Chapters 3 and 4), as they can assist designers in considering the consequences of product misuse (as it is highly likely that this is the cause of many of the accidents).

Product manufacturers already take this issue very seriously (Chapters 5, 6 and 8). Baber and Mirza (Chapter 5) show that most manufacturers' analyses tend to be investigations after accidents to determine if the accident was caused by a design fault. This is a reactive approach. A proactive approach may yield greater benefits, as suggested by BeSafe and TAFEI. If designers are able to reduce the frequency and magnitude of errors and accidents associated with their products, it could have a beneficial effect upon the sales of their products.

Interventions to reduce the incidence of error could include a range of strategies such as product redesign, labelling, packaging and modification of instructions. It is unlikely that any of these strategies could be implemented independently of the others. I was recently asked to comment on an instruction manual for a telecommunications product. I expressed my surprise at the request, that the client organisation thought that it was possible to assess the effectiveness of the manual independently of the product. Do users use the manual independently of the product? It was my assertion that it was the relationship between the manual and the product that must give rise to the user difficulties encountered and the most effective strategy to discover the cause of these problems would be to observe and interview a representative group of users as well as assessing the functional literacy of the manual. Therefore all interventions should be holistic.

## 15.6 'Prediction Is Difficult, Especially of the Future'*

We may speculate what the future holds for consumer products. Innovations in technology and design seem to be perpetually creating consumer demand for new products. I am of the generation that can remember black and white television, top-loading washing machines and reel-to-reel tapes. Within my lifetime I have seen the advent of colour television, video cassette recorders, compact discs, front-loading automatic washing machines, microwave ovens and dishwashers. I have no reason to believe that the rate of technological progress is going to slow down. Indeed, at a workshop on smart materials, organised by the Netherlands Design Institute in 1996, I was left with the distinct impression that technological progress would increase at an unprecedented rate. Futurologists at British Telecom are predicting the rate of technological progress to be somewhere in the region of one thousand- to a million-fold over the next 25 years. This projection is based on an expected increase in computing power and associated technological progress.

* Niels Bohr

I expect that this will mean *smarter* consumer products. To some extent we already have smart products, for example VCRs that record programmes whilst we are out, kettles that switch themselves off, microwaves that can cook food automatically and lights that come on when it is dark. The idea of smart products is that they are imbued with some degree of intelligence. Smart products are adaptive to the environment, the task requirements and the user. Truly smart products would not only be able to sense changes in the environment and respond accordingly, but also be able to learn about the effectiveness of the response via feedback. It might be argued that the smart concept would make the designers' task easier, as they do not have to define the operating envelope for the product (a requirement of ISO 9241) as the product adapts to changes in the environment.

These ideas might seem far-fetched, but product designers are already beginning to consider smart product concepts. Philips, for example, in its Visions of the Future exhibition considered over 300 possible product concepts in the four categories of personal, public, domestic and mobile. An example of a personal item was multimedia clothing that integrates information, communication and entertainment functions. The intelligent garbage can, which optimises waste disposal by sorting, compacting and removing odours was an example of a domestic product. Other examples include pull-out travel guides with flexible screens for interactive travel guide information (a mobile product) and rentable multimedia devices for special occasions (a public product).

Given that technological development is likely to continue unabated, the challenge for human factors is to develop robust methodologies for analysing human/product interactions as well as developing a greater understanding of the psychology of the product user. This is likely to be a more profitable route for research and development effort, rather than to be engaged in a continuous cycle of product testing each time that a new product is developed. We can be very certain that human evolution is going to remain fairly static for the coming century when compared with technological evolution. Therefore, if our efforts are focused on developing user-centred theories, principles, methods, guidelines and data, human factors will make an enormous contribution to the design of consumer products in the future.

# Human factors and ergonomics society contacts worldwide[1,2]

José Eduardo Grosso
Palacios 2091
Hurlingham - 1686
Buenos Aires
ARGENTINA
Tel.: +54 1 452 5980

Ms Margot Lynch
Ergonomics Society of Australia
Canberra Business Center
Bradfield St (cnr Melba)
Downer ACT 2602
AUSTRALIA
Tel.: +61 6 242 1951
Fax: +61 6 241 2554
Email: esa@ozemail.com.au

Dr Andreas Weiss
Institut F. Arbeits U.
Betriebswissenschaften Der Tu-Wien
Theresianumgasse 27
A1040, Vienna
AUSTRIA
Tel.: +43 1 658 315
Fax: +43 1 504 7146

Dr Dirk Delaruelle
Renault ind. Belgium SA
Schaarbeeklei 609
1800 Vilvoorde
BELGIUM

Professor Leila Gontijo
Univ. Federal de Santa Catarina
Caixa postal 476
88049 Florianopolis Santa Catarina
BRAZIL
Tel.: +55 48 231 9428
Fax: +55 48 231 9770

Dr Alexander Nikov
Technical University of Sofia
Dept of Human Sciences
PO Box 41
BG-1612 Sofia
BULGARIA
Tel.: +359 2 512 909
Fax: +359 2 512 909
Email: Nikov@vmei.acad.bg

[1] Addresses are listed in order of country.
[2] List kindly supplied by Steve Konz, Editor, *Ergonomics International*.

Mr Peter Fletcher
Executive Manager, HFAC/ACE
6519B Mississuaga Road
Mississuaga
Ontario L5N 1A6
CANADA
Tel.: +1 416 567 7193
Fax: +1 416 567 7191

Nora Zelaya
Centro de Ergonomia
Medicina y Salud Ocupacional – CEMSO
Apoquindo 4100, Ofic. 1001
Las Condes
Santiago
CHILE
Tel.:+56 2 220 0966
Fax:+56 2 220 0966

Professor Runbai Wei
Shanghai Municipal Education Comm.
505 Shaanxi Road (N)
Shanghai 200041
CHINA
Tel.: +86 21 625 63 010
Fax: +86 21 625 50 026

Margarita Gonzalez de Uribe
Facultad Terapia Ocupacional
Carrera 30 No. 152-40
Santafe de Bogota DC
COLOMBIA
Tel.: +57 216 97 28 - 216 94 14
Fax: +57 614 13 90
Email: magonza@ibm.net

ILO Regional Office for Africa
01 B.P.3960
Abidjan 01
COTE d'IVOIRE

Ms D. Maslic Sersic
Salajeva 3
41000 Zagreb
CROATIA

Professor Ricardo Montero Martinez
Facultad de Ingeniera Industrial
ISPJAE, Marianao
Ciudad Havana
CUBA
Fax: +53 7 336 075

Dr Jaroslav Formanek
National Institute of Public Health
Srobarova 48
100 42 Prague 10
CZECH REPUBLIC
Tel.: +42 2 670 826 25
Fax: +42 2 673 111 88

Mette Elise Larsen
BST Kobenhavn Vest APS
Hvidovrevvej 80, 2.Sal
DK-2610 Rodovre
DENMARK
Tel.: +45 367 2 1310

Dr Moises Castro Soo, President
Soc. Ecuatroriana de Salud Occu
Ranalcaxar, 235 y L, de Guraycos
PO Box 7015
Guayaquil
ECUADOR
Tel.: +59 3 433 0706
Fax: +59 3 580 189

Dr Mohammed El-Nawawi
Ind. Engg and Biotechnology
Al-Azhar University
Cairo
EGYPT

The Hon. Gen. Secretary
Devonshire House
Devonshire Square
Loughborough
Leicestershire LE11 3DW
ENGLAND
Tel.: +44 1 509 234 904
Fax: +44 1 509 234 904
Email: ergsoc@ergonomics.org.uk

Mr O. Kristjuhan
Tallin Technical University
Ehitjate Fee 5
EE0108 Tallin
ESTONIA

Dr Hannu Stalhammar
Finish Ergonomics Society
Ministry of Labour
Box 524
00101 Helsinki
FINLAND
Tel.: +358 0 18 56 79 58
Fax: +358 0 18 56 79 50
Email:
hannu.stalhammar@pt2.tempo.mailnet.fi

Dr S. Bogopolsky
European Soc. Dental Ergonomics
17, Avenue d'Argenteuil
F-92600 Asnieres sur Seine
FRANCE
Tel.: +33 47 93 0272
Fax: +33 47 93 0527

Prof. Dr W. Laurig
Institute für Arbeitsphysiologie
Universität Dortmund
Ardeystr. 67
D-44139 Dortmund H
GERMANY
Tel.: +49 231 108 4361
Fax: +49 231 108 4402
Email: laurig@arb-phys.uni-dortmund.de

Mr Ilias Banoutsos
Ergonomia Ltd
3rd Septembriou 77
Athens 10434
GREECE
Tel.: +30 1 82 28 888

Professor R. Gooonetilleke
Dept of IE
Hong Kong Univ. of Science/Technology
Clear Water Bay
Kowloon
HONG KONG
Tel.: +852 2358 7109
Fax: +852 2358 0062
Email: ravindra@usthk.ust.hk

Dr Krisztina Lakatos, Head
OMFB-IFETI
PO Box 565
H-1374 Budapest
HUNGARY
Tel.: +36 1 266 0408
Fax: +36 1 266 0469
Email: antalovits@erg.bme.hu

Professor Dr R. N. Sen
HB260, Sector 3
Salt Lake City
Calcutta 700091
INDIA
Tel.: +91 33 350 1397/6387
Fax: +91 33 241 3222
Email: rnsen@cubmb.ernet.in

Professor Adyana Manuaba
Head, Dept of Physiology
School of Medicine
University of Odayana
Jalan P.B. Sudirman
Denpasar 80232 A
INDONESIA
Tel.: +62 0361 226 132
Fax: +62 0361 237 614
Email: adman@denpasar.wasantara.net.10

Dr Alireza Choobineh
College of Public Health
Shiraz University of Medical Sciences
PO Box 111
71645 Shiraz
IRAN
Tel.: +071 769 111 x112
Fax: +071 760 225

Steve Chan
Industrial Design
National College of Art and Design
100 Thomas Street
Dublin 8
IRELAND
Tel.: +353 1 671 1377
Fax: +353 1 677 8468

Dr Issachar Gilad
Technion-Israel Institute of Technology
Faculty of Ind. Engg and Mgmt
Technion City
Haifa 32000
ISRAEL
Tel.: +972 4 294 434, 294 451
Fax: +972 4 235 194
Email: igilad@ie.technion.ac.il

Prof. Dott. Alfredo Bianchi
Univ. di Catania
Cattedra di Farmacologia Medica
95125 Catania
ITALY
Tel.: +39 95 222 297
Fax: +39 95 222 297
Email: bianchi@ictuniv.unict.it

Professor Hideo Nakata
Secretariat of Human Ergology Soc.
c/o Inst. for Special Education
University of Tsukuba
Tsukuba-shi 305
JAPAN
Tel.: +81 298 53 6748
Fax: +81 298 53 6748
Email:
    h-nakata@ningen.human.tsukuba.ac.jp

Secretary
Ergonomics Society of Korea
Dept of Industrial Engineering
Korea University
Anamdong 5-1
Sungbukgu
Seoul 136-701
KOREA
Tel.: +82 2 925 5038
Fax: +82 2 929 5888

Dr Sergei Yurov
Bikernieku iela 29-23
Riga, LV-1039
LATVIA
Tel.: +371 2 248 601
Fax: +371 7 139 163
Email: amber@carry.neonet.lv

Professor Romulaldas Kurila
Kaunas Technical University
Darbo
233036 Kaunas
LITHUANIA
Tel.: +370 7 205 362
Fax: +370 7 208 757

Dr Guat-Lin, Evelyn Tan
Secretary-Treasurer SEAES
School of Housing, Building and Planning
Universiti Sains Malaysia
11800 Minden
Penang
MALAYSIA
Tel.: +60 4 657 7888 ext. 3972
Fax: +60 4 657 1526
Email: lintan@lintan.pc.my

Dr Enrique Bonilla
Calzada del Hueso no. 1100
Col. Villa Quietud, CP 04960
Coyoacan
MEXICO DF
Tel.: +52 724 5140
Fax: +52 723 5483
Email: bono33@cueyatl.uam.mx

Secretariat, Dutch Ergonomics Society
PO Box 84106
2508 AC Den Haag
THE NETHERLANDS
Tel.: +31 70 338 3710
Fax: +31 70 351 2620

Wallace Simmers
New Zealand Ergonomics Society
33 Monaghan Avenue
Karori
Wellington
NEW ZEALAND
Tel.: +64 44 76 9073
Fax: +64 44 76 4789

Mr Enefiok Udo
Dept of Forestry and Wildlife
University of Uyo
UYO
Akwa Ibom State
NIGERIA

Kirsti Vandraas
Secretary NES
H+G A/S, PB 5055, Majorstua
0301 Oslo
NORWAY
Tel.: +47 22 595 913
Fax: +47 22 595 959

ILO Regional Office for Latin America
    and the Caribbean
Apartado Postal 3638
Lima 1
PERU

Assoc. Prof. Dr Clarissa Rubio
Dept. of Sociology
College of Social Science and Philosophy
University of the Philippines
Diliman
Quezon City 1101
PHILIPPINES

Dr Halina Cwirko, MD
National Labour Inspectorate
Krucza st 38/42
00-512 Warsaw
POLAND
Tel.: +48 22 11 011 ext 546
Fax: +48 26 28 4113

Graca Gomes Pereira
Estrada da Costa
1499 Lisboa Codex
PORTUGAL
Tel.: +351 14 196 777
Fax: +351 14 151 248

Professor Andrey Redman
Voinovastr. 9, Ap 1
191187 St Petersburg
RUSSIA
Tel.: +78 12 234 8960
Fax: +78 12 314 3360
Email: 100303.2047@compuserve.com

Ergonomics Society of FR Yugoslavia
Lola Corp.
Bulevar Revolucije 84
11000 Belgrade
FR Yugoslavia
SERBIA
Tel.: +38 11 457 390
Fax: +38 11 457 390

Ms Linda Herman
Chair, Ergonomic Society of Singapore
Info. Tech. Inst.
11 Science Park Road
Singapore Science Park II
SINGAPORE 0511
Tel.: +65 770 5916
Fax: +65 779 1827
Email: 100255.323@compuserve.com

Ms Ing. Andrea Holkova
c/o Faculty of Mat. Sc. and Tech
STU Paulinska (st) 16
917 24 TRNAVA
SLOVAK REPUBLIC
Tel.: +42 805 226 36
Fax: +42 405 277 31

Professor Patricia Scott
Dept of Human Movement Studies
Rhodes University
PO Box 94
Grahamstown, 6140
SOUTH AFRICA
Tel.: +27 461 318 111
Fax: +27 461 250 49
Email: hmps@kudu.ru.ac.za

D. Miguel Angel de la Inglesia Perealta
Ergogroup S.L., Sabino Arana
8 (Mod. 4). 48103 Bilbao
SPAIN
Tel.: +34 94 427 81 60
Fax: +34 94 427 80 50

Dr Moneim Attia
Inter. Heat Stress Research
Dept of Physiology
Faculty of Medicine
PO Box 102
SUDAN

Marianne Karlsson
Dept of Consumer Technology
Chalmers University of Technology
S412 96 Gothenberg
SWEDEN
Tel.: +46 31 772 1108
Fax: +46 31 772 1111
Email: mak@mot.chalmers.se

Professor Helmut Krueger
Dept of Hygiene and Applied Physiology
Swiss Fed. Inst. of Tech. Zurich (ETH)
Clausiusstrasse 25
ETH Zentrum
CH - 8092 Zurich
SWITZERLAND
Tel.: +41 16 323 973 Ext 72
Fax: +41 16 321 173
Email: krueger@iha.bepr.ethz.ch

Dr Rungtai Lin
President, Mingchi Institute of
    Technology
84, Gungjuan Road
Taishan
Taipei
TAIWAN 243
Tel.: +886 2 901 4490
Fax: +886 2 901 1914
Email: rtlin@cguaplo.cgu.edu.tw

Dr Kitti Intaranont
Assoc. Dean of Engg Grad. Studies
Faculty of Engineering
Chulalongkorn University
Bangkok 10330
THAILAND
Tel.: +66 2 218 6410
Fax: +66 2 253 6161
Email: fengkit@chulkn.chula.ac.th

Professor Ahmet Ozok
Tech. Univ. of Istanbul
ITU Isletme Facultesi
80680 Besiktas
Istanbul
TURKEY
Fax: +90 212 240 7260
Email: isozok@tritu.bitnet

Professor Akiva Ashorov
Vice President, UEA
PO Box 9149
310058 Kharkov
UKRAINE
Tel.: +75 72 456 343
Fax: +75 72 456 343

Dr Raul D. Baranano
Chana 2393 esq. Br. Artigas
CP:11200, Montevideo
URUGUAY
Tel.: +598 249 7893
Fax: +598 241 8208

Lynn Strother
Executive Administrator
Human Factors and Ergonomic Society
Box 1369
Santa Monica
CA 90406
USA
Tel.: +1 310 394 1811
Fax: +1 310 394 2410
Email: 70732.2420@compuserve.com

Dr Nguen Ngoc Nga
Deputy Director
National Institute of Occupational Health
1B Yersin
Hanoi
VIETNAM

# Author Index

# Subject Index